Biological Metaphor
and Cladistic Classification

Biological Metaphor and Cladistic Classification

An Interdisciplinary Perspective

Edited by

Henry M. Hoenigswald
and Linda F. Wiener

upp

University of Pennsylvania Press

Philadelphia   1987

Library of Congress Cataloging-in-Publication Data

Biological metaphor and cladistic classification.

   Papers from a symposium on Biological Metaphor Outside Biology, held Mar. 4–5, 1982 and
an Interdisciplinary Round-Table on Cladistics and Other Graph Theoretical Representations,
held Apr. 28–29, 1983, both at the American Philosophical Society's Library in Philadelphia.
   Includes bibliographies and index.
   1. Biology—Classification—Congresses.   2. Cladistic analysis—Congresses.
3. Comparative linguistics—Congresses.   4. Historical linguistics—Congresses.
I. Heonigswald, Henry M., 1915–   II. Wiener, Linda F.   III. Interdisciplinary
Round-Table on Cladistics and Other Graph Theoretical Representations (1983 : American
Philosophical Society Library)
QH83.B575   1987   574'.012   87-5019
ISBN 0-8122-8014-8

CONTENTS

Contributors   vii

Preface   ix

Part One: Historical Perspectives   1

1. Biological Analogy in the Study of Languages
    Before the Advent of Comparative Grammar   3
    W. Keith Percival

2. The Life and Growth of Language: Metaphors in
    Biology and Linguistics   39
    Rulon S. Wells

3. "Organic" and "Organism" in Franz Bopp   81
    Anna Morpurgo Davies

4. On Schleicher and Trees   109
    Konrad Koerner

5. A Legal Point   115
    Boyd H. Davis

6. Haeckel's Variations on Darwin   123
    Jane M. Oppenheimer

Part Two: Methodology   137

7. Cladistic and Paleobotanical Approaches to Plant
    Phylogeny   139
    Peter R. Crane and Christopher R. Hill

8. Pattern and Process: Phylogenetic Reconstruction in
    Botany   155
    Peter F. Stevens

9. Characters and Cladograms: Examples from Zoological Systematics   181
   Michael J. Novacek
10. Reconstructing Genetic and Linguistic Trees: Phenetic and Cladistic Approaches   193
   Maryellen Ruvolo
11. Of Phonetics and Genetics: A Comparison of Classification in Linguistic and Organic Systems   217
   Linda F. Wiener
12. The Upside-down Cladogram: Problems in Manuscript Affiliation   227
   H. Don Cameron
13. Representing Language Relationships   243
   William S.-Y. Wang
14. Language Family Trees: Topological and Metrical   257
   Henry M. Hoenigswald
15. Computational Complexity and Cladistics   269
   David Sankoff

Index   281

CONTRIBUTORS

H. Don Cameron is Professor of Greek and Latin at the University of Michigan, where he also holds an appointment as curator in the Museum of Zoology. His zoological interests are in diptera and arachnida. He has published on Greek tragedy and the history of zoological nomenclature.

Peter R. Crane is Associate Curator of Paleobotany at the Field Museum of Natural History in Chicago, and is a specialist in the origin, early fossil history, and evolution of angiosperms. He has been concerned with clarifying the role of paleobotany in attempts to elucidate plant phylogeny and evolution.

Anna Morpurgo Davies, FBA, is Professor of Comparative Philology at Oxford University and a Fellow of Somerville College. She is active in Mycenaean and Anatolian studies, and has written on the history of linguistics.

Boyd H. Davis is Professor of English at the University of North Carolina at Charlotte. She is currently editing and contextualizing the manuscripts of Ferdinand de Saussure in the Houghton Library at Harvard University.

Christopher R. Hill is Curator of Palaeobotany at the British Museum (Natural History), London. His research has focused primarily on the evolution of cycads and marattialean ferns. He has been concerned with clarifying the role of paleobotany in attempts to elucidate plant phylogeny and evolution.

Henry M. Hoenigswald is Emeritus Professor of Linguistics at the University of Pennsylvania. His publications include *Language Change and Linguistic Reconstruction* and *Studies in Formal Historical Linguistics.*

E. F. Konrad Koerner is Professor of Linguistics at the University of Ottawa. He is the Editor of *Historiographia Linguistica* and heads the publishing program Amsterdam Studies in the Theory and History of Linguistic Science.

Michael Novacek is Chairman and Associate Curator of the Department of Vertebrate Paleontology at the American Museum of Natural History. He is interested in the early phylogeny of the major groups of mammals. He also conducts field projects in mammalian paleontology in the United States, Mexico, and South America.

Jane M. Oppenheimer is Professor Emeritus of Biology and History of Science at Bryn Mawr College. Embryology and its history are her special interests. Most recently (1986) she edited the first English translation of K. E. von Baer's *Autobiography.*

W. Keith Percival received his Ph.D. from Yale University in 1964, and has taught at the University of Kansas since 1969. He specializes in the history of linguistics, particularly the development of grammatical theory from the fourteenth to the sixteenth centuries. He has recently contributed chapters to the *Cambridge History of Later Medieval Philosophy* and to *Renaissance Eloquence.*

Maryellen Ruvolo is a molecular biologist at the Harvard Medical School, working on the evolution of and regulation of the immune system genes. Her graduate work in the Anthropology Department at Harvard University was done on the speciation of the African Guenon monkeys as traced in their blood proteins.

David Sankoff received his Ph.D. in mathematics from McGill University and has been a member of the Center for Mathematical Research at the University of Montreal since 1969. He has worked extensively on mathematical models, statistics, and computational methods useful in linguistics, biology, and other fields.

Peter F. Stevens is Professor of Biology and Curator of the Arnold Arboretum and Gray Herbarium in Cambridge, Massachusetts. His current interests include the development of the "natural" system in botany (especially between 1760 and 1860), monographic and phylogenetic studies in the Guttiferae and Ericaceae, and the theory of systematics.

William S.-Y. Wang received his Ph.D. from the University of Michigan in 1960. He is Professor of Linguistics at the University of California at Berkeley. He has written extensively for the *Journal of Chinese Linguistics,* on phonetics, and on various aspects of language evolution.

Rulon S. Wells is Professor of Philosophy and Linguistics at Yale University. He is a past president of the Linguistic Society of America.

Linda Wiener received her Ph.D. in entomology from the University of Wisconsin. She is on the faculty of St. John's College in Santa Fe, New Mexico.

PREFACE

Biological metaphor has been utilized in many fields for both artistic and heuristic purposes. The dynamic aspects of biological systems, the birth, growth, decay, and death of individual organisms, the change of one form into another over time, and the appearance and extinction of species are all compelling images.

Reconstructing the intellectual history of the use of biological metaphor in linguistics and other fields is a pertinent question from several points of view. When and why were these metaphors chosen? Were they truly heuristic or simply attention-getting devices? In what ways do the biological metaphors accurately reflect the linguistic problems to which they have been applied? How has the use of a metaphor changed over time?

One popular metaphor is the use of trees in classification. These are used in stemmatics, linguistics, and biology. All these disciplines construct trees that are meant to mirror relationships among various groups and to reflect actual changes that have occurred over time. What are the techniques, the evidence, and the problems associated with constructing tree classifications? What are the alternatives?

When biological taxonomists and historical linguists are brought together and begin to talk about their respective disciplines they are often amazed at the number of issues and problems they share. These include the problems of sub-grouping, character selection and weighting, identification of presumed ancestors, geographic variation, and methods for representing both synchronic and diachronic relationships.

Although biological classification and historical linguistics both have histories extending back into classical times, the "sciences" of biological taxonomy and historical linguistics both emerged as professional disciplines in nineteenth-century Europe. This was an exciting time in both fields. Evolutionary biology was founded and expounded by Darwin, Wallace, Huxley, and Haeckel. Cuvier founded the field of comparative anatomy; von Baer was elucidating the mysteries of embryology and development. In linguistics Bopp, Grimm, Schleicher, Brugmann, and others were discovering relationships between languages and gathering evidence which revealed phonological laws and allowed lines of descent to be traced. Did evolutionary biology and the tree metaphor influence the development of historical linguistics? Were there influences from other fields as well? A variety of opinions is expressed in the historical perspectives section of this volume.

W. K. Percival discusses the concepts of organic and inorganic, and the family tree metaphor, as they occur in ancient Greece and Rome, through the Hebrew and Arabic grammarians of the Middle Ages, up to the European linguists of the nineteenth century. Although other papers in this volume stress a close connection between nineteenth-century biological thinking and the founding of modern historical linguistics, Percival highlights the two and a half centuries of genealogical thinking about languages which preceded this.

In contrast, A. Morpurgo Davies shows how use of biological metaphor actually influenced the technical work of Franz Bopp. Bopp saw language as an organism which grew, decayed, changed and passed itself on to new generations. Although this intellectual framework was rather ill-defined, Bopp was not averse to using it to make decisions concerning the reconstruction of Indo-European.

K. Koerner explicitly discusses contemporary influences on Schleicher, a pioneer of historical reconstruction in linguistics. He argues that Schleicher's tree diagrams were strongly influenced by tree classifications in the natural sciences as well as by tree construction in stemmatics (the study of manuscript affiliations).

The issue of tree iconography is raised by J. Oppenheimer in her paper on Haeckel, a popularizer of Darwin's evolutionary views. She shows how Haeckel developed this metaphor in both written and diagrammatic fashion.

Finally, R. Wells expounds on the whole problem of using biological metaphors in linguistics. Mentioning that these metaphors have been used more for their affective than for their cognitive meanings, he claims that they were eventually abandoned by professional linguists in the nineteenth century for this reason. Consider that the equation of languages with biological organisms may be extended to growth, death, and reproduction—but certainly does not extend to a need for specific temperature ranges or the stages of puberty or menopause or the like. Only part of the analogy is potentially heuristic, and many possible analogies are unfruitful.

It seems clear that the intellectual history of historical linguistics is quite complex and has drawn on many lines of influence to reach its present state. Biological metaphor has been heuristic in some ways, but misleading in other respects. From this historical perspective, it is interesting to look at current methods of classification in these fields.

For the past ten years or so there has been a kind of "revolution" going on among systematists in the biological community. This revolution is termed "cladistics" and is claimed to be both a more scientific and a more accurate method for constructing phylogenies and representing information about relationships among groups of organisms.

Cladistic classification involves constructing a tree in which branch points are based on possession of derived or more recent characters as distinguished from primitive characters. Other competing methods of classification differ in that they use measures of over-all similarity without distinguishing derived from primitive characters (phenetics) or that they consider information additional to possession of shared derived characters, such as ecological niche or amount of change in a taxon since branching (evolutionary taxonomy).

The controversy surrounding the cladistics revolution has been constructive in many ways. Taxonomy has often been considered a dull and unproductive specialty among biologists who study more "exciting" things. Cladistics has had the happy effect of bringing taxonomy out of dark, musty museums and into the forefront of biological research. Important questions have been raised about how to transform taxonomy from an art into a science with rigorously defined procedures, testable hypotheses, and repeatable results. Not much public soul-searching seems to be going on in the linguistics community today, although linguists are certainly aware of the problems associated with constructing genealogical trees. As we shall see, different methodologies produce different taxonomic trees, and everyone who uses supposed historical relationships between languages or organisms as the taking off point for other comparisons should be aware of the issues involved.

It often comes as a surprise to biologists that the predominant method in linguistic classification has been cladistic for well over one hundred years. However, whereas trees based explicitly on measures of overall similarity preceded cladistic methods in biology, such a procedure has emerged rather recently in linguistics, under the name of lexicostatistics or glottochronology.

The papers in the methodology section discuss the intricacies of various methods of classification. The papers range from those which broadly compare to those which focus on specific problems in particular disciplines. It will become obvious that the problems connected with historical reconstruction in these fields overlap quite a bit. For the uninitiated, good introductions to the basic methodologies can be found in H. D. Cameron's paper on manuscript affiliation, H. M. Hoenigswald's paper on language family trees, and M. Ruvolo's paper on phenetic and cladistic approaches to biological data. L. Wiener's paper gives a comparative overview of philosophical, methodological, and practical issues in historical reconstruction in biology and linguistics.

While stemmatics has always used the cladistic approach to family tree construction (the shared error defines the branch point), biology and linguistics have developed both cladistic and phenetic methods. At present, cladistic methods are largely preferred in both fields, for reasons M. Ruvolo discusses. Phenetic methods have often seemed attractive because a great deal of data could be handled relatively quickly and easily, and it appeared that fewer arbitrary decisions needed to be made. D. Sankoff discusses the problem of dealing with large data sets and gives insight into the computational difficulties of finding the best cladistic tree. He shows that making a few a priori assumptions can greatly facilitate this task and reduce the problem to the point where computer analysis to find the most parsimonious tree becomes practical.

From this background, we can see how more specific issues are dealt with. W. Wang's paper on language relationships brings up many problems which are echoed in the biological papers. These include character weighting (determining in the face of conflicting evidence which characters are more important for tree construction), hybrid taxa, the possibility of changes reversing themselves, binary vs. multifurcating trees, the use of nonbinary characters (which cannot be scored with a simple + or −), and the use of transformation series.

P. F. Stevens' paper on botanical systematics discusses many of these problems in the context of plant classification. Plants share with languages a severe problem with homoplasy or convergent development. Plants, as a rule, hybridize far more freely than animals and so both plant and language lineages present the problem of having to fit a hierarchal diagram to a group of taxa which have very likely evolved in a reticulate manner.

M. Novacek emphasizes problems in zoological tree construction. He discusses how characters are defined, chosen, and weighted, and how to decide which character of a pair is older. This problem of making polarity decisions is not considered a major difficulty in linguistics. Phonological change is presumed to be one-way, thereby making the polarity of a change obvious. Also, historical evidence is treated differently in these fields. As P. Crane and C. Hill make clear in their paper on paleobotanical evidence, one must use great caution in evaluating the historical record, and cannot assume that a fossil organism has the more primitive character. Linguists, on the other hand, give historical records a very special status in historical reconstruction and weight such evidence heavily.

Many of the authors in this section stepped out of their fields of specialty to comment on and compare their fields to others. Thus, M. Ruvolo shows how certain techniques used to evaluate biochemical data could be used (or abused) with lexicostatistical data. P. Stevens compares the task of reconstruction in botany, zoology, and linguistics, and H. D. Cameron philosophizes about the iconography of tree diagrams in stemmatics, linguistics, and biology.

In all, the papers in this section should give a good introduction to the special methods and problems associated with classification in these fields and give the reader a feeling for the common problems as well. It is apparent that intellectual history, the nature of a system, and the nature of the evidence all influence the methods currently employed by practitioners in these fields. Taken together, the papers in this volume should give an idea of the complexity of historical and contemporary problems in this area and even suggest some possible solutions. This meeting of disciplines reveals that seemingly disparate fields such as biology, linguistics, and stemmatics do share common problems, and we hope that an interdisciplinary perspective may help everyone take a fresh look at these issues in order to clarify problems and develop better and more accurate methodologies. We hope that this volume will stimulate the reader to make his or her own comparisons.

After a preparatory period in 1981, during which a faculty seminar on Comparison in the Humanities and in Science met at frequent intervals at the University of Pennsylvania, a symposium on Biological Metaphor Outside Biology was convened on March 4 and 5, 1982. This was followed on April 28 and 29, 1983, by an Interdisciplinary Round-Table on Cladistics and Other Graph Theoretical Representations. This volume mainly contains contributions to the 1982 and 1983 meetings.

All participants in these meetings were asked to submit their contributions for publication. We regret that some of the papers presented could not be included here for one reason or another. Many participants who did not present papers spoke in the discussions at both meetings, and we wish to thank those individuals for their contributions.

We are deeply indebted to the munificence of the Dean of the School of Arts and Sciences at the University of Pennsylvania and to the School's Humanities Coordinating Committee for supporting all these activities and this publication, and to the Librarian of the American Philosophical Society, Dr. Edward C. Carter II, for inviting both the symposium and the round-table to meet in the splendid surroundings of the Society's Library in Philadelphia. We are also most grateful to many colleagues who have advised and aided us, none more than Bentley Glass, Henry Hiż, Alan E. Mann, and Jane M. Oppenheimer.

HENRY M. HOENIGSWALD
LINDA WIENER

PART ONE

Historical Perspectives

1

Biological Analogy in the Study of Language Before the Advent of Comparative Grammar

W. KEITH PERCIVAL

The focus of this chapter is the use of biological analogies and metaphors by grammarians and linguists prior to the founding of comparative linguistics as an academic discipline in the early nineteenth century. However, one cannot reasonably discuss pre-1800 developments without relating them to what happened subsequently, so I shall begin by discussing the nineteenth century. What then follows will be a retrospective survey. The concepts we usually regard as having played a role in nineteenth-century linguistics will be the point of departure, and I shall follow them back in time. This will be neither pure history nor pure theory, for both theoretical and historical issues are intertwined here, but it is hoped that the historical evidence presented in this chapter will shed light on the theoretical aspects of the problems we are facing.

Perhaps the most vexing issue is a fundamental one, namely, the extent to which we are justified in approaching the study of language by way of metaphors and analogies. At least one famous linguist, Ferdinand de Saussure, adopted a radical position in this regard. "We are firmly convinced," he once wrote in an unpublished article, "that anybody who sets foot in the realm of language may consider himself abandoned by all the analogies of heaven and earth." [1]

Few linguists today share this viewpoint, and the average historian of linguistics would probably concur, pointing out that the study of a phenomenon as protean as human speech is bound to be directed into analogical and metaphorical channels by a multitude of practical as well as theoretical considerations. Traditional grammar, as conducted in the West, is a case in point. From Hellenistic times until recently, grammarians regarded their discipline as an *art,* that is, an organized system of prescriptions. To approach the study of languages this way is to indulge in a vast metaphor equating language and writing with acquired skills like syllogistic reasoning. Present-day linguists regard the analogy as invalid, and perhaps the enterprise of traditional grammar was indeed fundamentally flawed. In this connection, John Lyons has spoken wryly of the "classical fallacy."[2]

I shall neither defend nor attack traditional grammar. The development of linguistic study in these peculiar channels is a historical fact, and without it linguists would not be where they are today. My task here will be to draw the historical map as carefully as possible, and only then to raise questions of justification or perhaps to defer to others more qualified than myself.

In most instances, the use of analogies and metaphors may have been forced on grammarians either by pragmatic factors such as pedagogy or by intellectual currents so pervasive that their underlying assumptions were taken for granted. One thinks, for example, of the devotion to the study of ancient literature in the case of the Greeks, or the preoccupation with religious texts and ritual in the case of the Hebrew and Sanskrit grammatical traditions.

A further complicating factor is the degree to which different scholarly disciplines, once well established, influence one another. In certain circumstances, concepts and techniques exert an appeal outside the disciplines in which they originated. This may happen in part because the practitioners of any academic specialty, no matter how well initiated, receive training in a restricted repertory of techniques and by that fact are at times hampered in tackling the very problems which their discipline exists in order to solve. After all, there is no *a priori* reason why all problems should respond to precisely those skills that the expert happens to have acquired during his professional initiation. Fortunately, few people are completely devoid of general intellectual interests, and most scientists and scholars acquire at least a smattering in other subjects before proceeding to concentrate on their own. It is perhaps for this reason that unsolved problems sometimes find unexpected solutions from scholars who happen to be equipped with techniques foreign to the discipline in which they were trained.

But let us turn to the main topic: the way grammarians and linguists have utilized concepts of a biological nature. My own explorations in this area have been hampered because the topic has hitherto attracted little attention from historians of linguistics. However, a good basis for study is provided by a number of fine treatments of the history of etymology and the history of theories about the origin of language. The earliest permanent contribution to the history of etymology was made by Laurenz Lersch almost one hundred and fifty years ago.[3] More discussion of the characteristics of ancient etymology can be found in a comprehensive survey of ancient linguistics written by the nineteenth-century linguist H. Steinthal.[4] There is also a stimulating treatise on the philosophy of language in the patristic and scholastic periods by Paolo Rotta, written in the early twentieth century, in which there is a discussion of Plato's *Cratylus*.[5] In the 1920s, Ernst Cassirer surveyed the history of the philosophy of language in the first volume of his *Philosophie der symbolischen Formen*.[6] More recently, the medieval historian Arno Borst published a multivolume study of theories relating to language diversity.[7] The most recent major contribution to the topic is a monograph by James Stam.[8] Finally, on the history of what linguists today call the "comparative method," a number of valuable studies have appeared.[9]

In general, however, the degree to which specifically biological analogies or terminology were in evidence at various periods in the history of linguistic studies is not dealt with in any detail in the books and articles I have mentioned. Since this chapter therefore cannot aspire to be more than a tentative sounding of the depths of our present ignorance in this area, there will be a number of loose ends to stimulate discussion and further study. To be sure that we are clear about the number of separate conceptual components making up the phenomenon we are to examine, let us first enumerate the major biological analogies that played a role in nineteenth-century linguistics.

There is, first, the notion of language relatedness. In nineteenth-century (as in present-day) linguistic terminology, languages are said to be related to one another if they are assumed to be later divergent continuations of a single earlier language, the latter being referred to as the "parent language" (Ger., *Ursprache*).[10] For instance, English and Swedish are related languages in this sense, and nineteenth-century linguists referred to the parent language of which English and Swedish are later divergent continuations as "Primitive Germanic" (Ger., *Urgermanisch*).

A set of related languages is called a language family or language stock (in the mid-nineteenth century August Schleicher used the terms *Sprachsippe* and *Sprachstamm*), and related languages are often referred to as daughter languages (Ger., *Tochtersprachen*), said to "descend" from the parent language (or "protolanguage," as we usually say today). Accordingly, two daughter languages descended from the same protolanguage may be called "sister" languages (Ger., *Schwestersprachen*).

When the process of differentiation repeats itself, it gives rise to what the French Indo-Europeanist Antoine Meillet called "intermediate" protolanguages (*langues communes intermédiaires*).[11] This makes it possible to display the relationships among the languages of a family in a branching diagram analogous to a genealogical tree (Ger., *Stammbaum* or *Stammtafel*) or a philologist's stemma of the Lachmannian variety.[12] This was done, for instance, by August Schleicher in the introductory chapter of his *Compendium der vergleichenden Grammatik,* first published in 1863.[13] It is significant, moreover, that Schleicher's diagrams were not mere logical schemata. As he himself explains, "The length of the lines indicates the amount of time which had elapsed and the distance between them degrees of relationship."[14] In this respect, Schleicher's trees differed from the traditional tree of Porphyry or Porphyrian scale, which was a purely logical method of exhibiting the series of subaltern genera to which a concept may be assigned.[15]

Using another genealogical analogy, nineteenth-century linguists spoke of the various "branches" of a language family (Ger., *Zweige*): the Germanic branch, the Slavic branch, and so on. The practice of arranging languages in genealogical trees was sharply criticized by a number of linguists in the nineteenth century.[16] Despite many cogent arguments, however, comparative linguists did not abandon the use of branching diagrams; in fact, the family tree has remained very much a part of linguistics to this day.

Second, in addition to language relationship, there was the important notion of relationship among words, or lexical cognation, "cognates" being words in related languages which could be shown to be later more or less divergent forms of a single word in the parent language, or as the *Oxford English Dictionary (OED)* put it, words "coming naturally from the same root, or representing the same original word, with differences due to subsequent separate phonetic development." French *bouche,* Spanish *boca,* and Italian *bocca,* all meaning "mouth," are cognates in this sense, each representing later altered continuations of the Latin *bucca* 'mouth.' Cognate words were therefore related to one another in much the same way as related languages. In fact, the terms "cognate" and "related" were often and still are sometimes used interchangeably. The word in the parent language was called "root," more precisely the "etymon," of the cognates in the descendant languages. Etymologically speaking, cognate is a genealogical term: Latin *cognatus* 'related by birth.'

Third, the concept of the root was utilized in the morphological analysis of individual languages, a root being "one of those ultimate elements of a language, that cannot be further analysed, and form the base of its vocabulary" (*OED*). In practice, this meant that roots were what traditional grammarians had called *primitive* words, as opposed to *derivatives.*

A number of other terms related in meaning to "root" were current among grammatical analysts in the nineteenth century: base, stem, and theme. Of these, theme already had a long history and originally referred to the nominative singular form of nouns and the first person singular present indicative form of verbs, which were the forms customarily listed in dictionaries of the two classical languages. The term seems to have been inherited from the Greek grammatical tradition (*thēma*).[17] Theme was also equated with root and widely used in Western grammars of Hebrew,[18] but its modern meaning (i.e., synonymous with stem) is no older than Georg Curtius, and hence from the mid-nineteenth century. Stem was originally used in the sense of "the primary word from which a derivative is formed" (*OED*). In its present-day meaning (the most recent edition of the *Concise Oxford Dictionary* defines it as "part of noun, verb, etc. to which case-endings etc. are added, part that appears or would originally appear unchanged throughout the cases and derivatives of a noun, persons of a tense, etc."), it seems to be a nineteenth-century creation. The term *Wortstamm* 'word stem' was used by Bopp, in addition to the more frequent *Grundform* 'base form'.[19] Stem was also still used in the same century to refer to the Semitic triliteral root.[20]

Finally, in addition to talking about language relationship and word relationship, nineteenth-century linguists often spoke of the grammar of a language as an "organism." However, it is interesting to note that syntax tended not to be regarded as part of the organism of a language. As Bopp once put it, "Inflections constitute the true organism of a language,"[21] and in a similar vein Jacob Grimm once referred parenthetically to syntax as "lying half outside grammar."[22] These organismic metaphors are already frequent in the writings of Friedrich Schlegel at the beginning of the nineteenth century, and in general a decided predilection for the organismic as against the mechanical is observable in belletristic literature as far back as the mid-eighteenth century.[23] The section of grammar dealing with word inflection and word formation was renamed morphology (it had been traditionally called *etymology*). The term "morphology" had been used earlier by biologists (see "morphology" in the *OED*).

Certain forms, sounds, or even letters were referred to by nineteenth-century linguists as "organic," while others were categorized as "inorganic." For instance, the final *e* of Modern English "these" was said to be organic because of the Middle English form *þise,* whereas the final *e* of "those" was inorganic because of Middle English *þas, þos* (these examples are cited under "Organic" in the *OED*). In practice, therefore, organic meant "reflecting the structure of the language at some earlier stage" or, perhaps more precisely, "reflecting the *original* structure of the language."

The logical consequence of these and other organismic parallels was clearly stated, for instance, by August Schleicher in his provocative pamphlet *Die Darwinsche Theorie und die Sprachwissenschaft*, first published in 1863. The distinctive feature of Schleicher's theory is that he regarded the analogies not as metaphorical *jeux de mots* but as realities. For him languages *were* organisms. As an illustration I quote the following passage from his *Darwinsche Theorie:*

Languages are natural organisms that arose independently of human volition, grew in accordance with definite laws, and developed and in turn grew old and died. They are also characterized by the set of phenomena we are accustomed to understand by the term "life." Accordingly, glottics, the science of language, is a natural science; its method is largely the same as that of the other natural sciences. This is why the study of Darwin's book . . . was bound to seem quite close to my area.[24]

Having reviewed the major biological analogies in use by linguists in the nineteenth century, let us now pose the following question: How far back in time were these analogies, or other analogies of the same general kind, in use, and what significance was attached to them? As we shall see, clear answers to these simple-looking questions are not easy to find.

Let me make one assumption at the outset: the notion of the structure or grammar of a language as an organism was peculiar to the nineteenth century, having been ushered in by the founders of comparative grammar (I am thinking here especially of Franz Bopp and the Schlegel brothers). The other biological concepts I have just mentioned already existed in ancient, medieval, and Renaissance thought. The conceptual basis was laid on the one hand in Greek and Roman antiquity and on the other hand in the biblical legends of the Flood and the Tower of Babel, and we observe a noticeable increase in the use of biological metaphors as we approach modern times. More specifically, one may distinguish two important watersheds since antiquity, the first being the sixteenth century and the second being the turn of the eighteenth and nineteenth centuries.

Anna Morpurgo Davies examines the second watershed in this volume (Chapter 3). Before I discuss the earlier one, let me digress on the notion of watersheds. In spite of appearances, nothing ever completely revolutionizes a discipline. If we want to understand the result of a conceptual revolution, we are well advised to investigate the system that preceded it. Much more survives in a discipline from each prerevolutionary stage than the practitioners of the discipline at the time are aware of and are willing to admit. As intellectual historians are fond of pointing out, there is a tendency to rewrite history after all revolutions. It is important, therefore, not to be unduly impressed by the professional rhetoric produced at vital junctures in the history of ideas. Hence, as revolutionary as the two watersheds I have hypothesized may have been from certain points of view, they necessarily left a great deal intact.

The sixteenth century, the locus of our first watershed, was the century in which there emerged in western Europe something one might call "general" linguistics. This development was stimulated by a number of factors, of which one of the most important was the newly instituted study of Hebrew and the other Semitic languages, namely, Arabic and Aramaic. The study of these languages introduced Latin-speaking scholars to a transparent case of a set of related languages. Native grammarians of these languages had long been aware of the relationship between them; indeed, the early grammars of Hebrew were influenced by and modeled on Arabic grammar, and many of them were written in Arabic.[25]

Before Westerners began to study Hebrew, that is, prior to the late fifteenth century, grammarians in western Europe taught Latin (and also, but to a limited extent, ancient Greek) and had little interest in the notion of language relationship. They also were not especially curious about the phenomenon of language diversity. In Renaissance Italy, for instance, grammar was still *by definition* Latin grammar, not general linguistics, even in the most liberal interpretation of that term. To take one example, the Latin grammars written by the humanists in the fifteenth century (e.g., the widely disseminated *Regulae grammaticales* by Guarino Veronese, 1374–1460), did not contain the word "Latin" in their titles. The first Latin grammar to have a title that included the word "Latin" was a work by Antonio de Nebrija: the *Introductiones Latinae explicitae,* first published in Salamanca in 1481. It is surely no accident that Nebrija was a student of Hebrew and also a vernacular grammarian; witness his famous *Gramatica sobre la lengua castellana* (Salamanca, 1492).

If we now go all the way back to Greek antiquity, what we observe is the development of a theory of etymology. This was an outgrowth of the Greek passion for word origins, already observable in the Homeric poems.[26] Perhaps interest in word history originated from the kind of folk history characteristic of so many early civilized societies in which history was largely genealogy, in particular the genealogy of the founding fathers of the states or communities in question. An important concept in this kind of speculation was the eponymous founder or hero, that is, the man, mythological being, or deity who gave his name to a city or other geographical or ethnic entity. For instance, it was generally assumed in antiquity that the Greeks (Hellenes) were named after Hellen, whose sons were Dorus, Xuthus (father of Ion), and Aeolus, who were in turn supposed to have founded the three basic Greek ethnic divisions recognized in antiquity: the Dorians, Ionians, and Aeolians. From that sort of ethnological speculation it is a small step to the giver of names in general (Gk., *onomatothetēs*).

Pythagoras seems to have been the first philosopher to give currency to the idea of the name-giver. The earliest copious source of information we have on this kind of speculation is Plato's *Cratylus*.[27] In that dialogue, the character Socrates develops an etymological theory in the course of cross-examining Hermogenes, who represents a crude conventionalist theory with regard to the relation between words and their referents. The opposing character, Cratylus, defends the notion that the relation between words and things in a natural one and that given the essence of a thing its name necessarily matches that essence. According to Cratylus, any name not related to its referent in that way would not be its *true* name. (Note that I attribute this theory not to Plato but to the character Socrates, because Plato himself seems to have had an ambivalent attitude to etymology. On the one hand, Plato denied that the essence of a thing could be arrived at by way of etymology, which is the main message of the dialogue, but on the other hand, he seems not to have been able to resist the lure of etymological speculation. On the whole, at least until recently, posterity took it for granted that Plato wholeheartedly supported the theory expounded by the character Socrates in the *Cratylus,* this interpretation being itself of great importance in the history of the philosophy of language. The person responsible for this misconception seems to have been Marsilio Ficino, the reviver of Platonism in Renaissance Italy, who published the first Latin translation of the dialogue together with a commentary; the first printed edition appeared in Florence in 1484.[28])

The theory that Socrates elaborates has the following features. First, he suggests that all words of contemporary Greek were ultimately derived from a set of primordial words (Plato's term is *prōta onomata* 'first words') by two main processes: phonetic degeneration or semantic change of individual primordial words, on the one hand, and compounding of two or more primordial words on the other. Thus, Socrates suggests (410B) that air may have been called *aēr* because it raises things from the earth (*hoti airei ta apo tēs gēs*). Some words have more extensive pedigrees: "The word *terpsis* ('delight') comes from *terpnon* ('delightful'), and *terpnon* in turn is from the creeping (*herpsis*) of the soul, which is so called because it resembles a breath (*pnoē*) and ought more appropriately to have been called *herpnoun,* but the word in course of time has been transformed into *terpnon*" (419D). *Mēkhanē* 'contrivance,' according to Socrates, came about because it means essentially *to anein epi poly* 'the act of accomplishing much,' the first part of *mēkhanē* coming from the word *mēkos* 'length,' which Socrates says is equivalent to *poly* 'much,' and the last part from *anein* 'to accomplish' (415A). The actual processes of phonetic degeneration involve removal and addition of sounds but are not further specified and classified by Plato.

Furthermore, the primordial words themselves were originally echoic, that is, imitative of their referents; in its earliest stage language faithfully mirrored reality. There is even an attempt by Socrates to assign semantic values to some of the sounds of Greek (beginning at 426C); for instance, rho is supposed to express movement, delta and tau are said to express binding and rest, and so forth. Socrates tacitly assumes that the number of primordial words was quite small and that this restricted set of elements gave rise to a much larger set of elements, namely, the whole vocabulary of contemporary Greek. Basic to the whole theory was the idea that the Greek language was simply the set of words used by Greeks. Plato had no concept of grammar.

But Plato was aware of linguistic change. In fact, he has Socrates say that change is so extensive that the primordial language would appear to his contemporaries like a foreign tongue (421D). There is, in addition, the supposition that the language has deteriorated in the interval, largely as a result of the love of euphony, and that we therefore need to go back to the primeval stage if we want to understand the true meanings of words. Socrates says to Hermogenes:

My dear friend, you fail to understand that the primeval words have already been buried by people who wanted to embellish them by adding and removing letters to make them sound better, and disfiguring them totally, either for aesthetic considerations or as a result of the passage of time. . . . My feeling is that things like this are done by people who have no thought for the truth while they are putting their mouths in shape. The result is that they insert many sounds into the original words until finally nobody can figure out what the words used to mean. . . . For if people are allowed to insert and remove whatever they please in words there will be nothing to prevent any word whatever from being adapted to refer to any object (414C–D).

This is an early testimony to the belief, which prevailed throughout antiquity, the Middle Ages, and the early modern period, that the contemporary language represents a long process of decay and degeneration and that only the primordial language has an inherent ontological justification.

Thus, the basic idea behind the etymological speculations that we witness in Plato's *Cratylus* was that the source of a word is not merely the historically earliest ascertainable form, but the actual *etymos logos* of the word, that is, its "true" meaning in the fullest sense of the term. From this it followed that etymology can provide us with correct definitions. Indeed, it was believed that a word cannot be adequately defined without recourse to etymology. In other words, etymological investigation was designed to be ontologically revealing; the etymologist was not simply looking for historical facts. This conception of etymology existed throughout antiquity, the Middle Ages, and the early modern period, and it still survives at times in the layman's view of etymology, but it is foreign to the aims of modern etymology as it has been conducted in the past hundred years or so by professional linguists.[29]

Another important fact is that etymology arose among the Greeks several centuries before a grammatical tradition took shape. If grammar had preceded etymology, as it perhaps did in India, the history of linguistics in the West might have been quite different. The separate status of grammar and etymology in antiquity meant that etymological speculations were conducted quite independently of morphological considerations. For instance, Chrysippus suggested that the Greek noun *aiōn* 'age,' 'period,' 'era' was derived from the phrase *aei ōn* 'being for ever.'[30] Similarly, in late antiquity the Latin word *testamentum* 'will,' 'testament' was said to come from *testatio mentis* 'the mind's witnessing' (Justinian, *Institutiones* 2.10). This is, of course, quite impossible. Morphologically speaking, the word *mens/mentis* 'mind' does not form part of *testamentum,* and hence the etymology has no basis whatever. It may seem incredible to us today that for so long few scholars thought to criticize the principle underlying etymologies of this type. In fact, a revulsion did not occur until the fifteenth century, when Lorenzo Valla (1407–1457) poured scorn on a number of etymologies from various ancient authors in book 6 of his widely disseminated stylistic manual *De elegantia linguae Latinae,* first published in the mid-1440s.[31]

After Plato, indeed after Aristotle, we find an etymological theory expounded by the Stoics which bears a striking resemblance to that of Plato's character Cratylus. Unfortunately, as is in general the case with Stoicism, the original documents have not come down to us, and we are reduced to reconstructing the theory from accounts in later writers. Two important sources are the extant portions of Varro's *De lingua Latina,* which dates from the middle of the first century B.C., and a treatise on logic traditionally ascribed to Augustine (A.D. 354–430).

By the time Varro was writing, a tradition of technical grammatical writing was developing among both the Greeks and the Romans. Varro himself wrote a grammatical textbook of Latin (it formed the first book of his encyclopedic *Disciplinarum libri IX*), but it was soon to be superseded by the work of Remmius Palaemon in the first century A.D. and has not come down to us.[32]

In Varro's *De lingua Latina,* which is not a grammatical textbook but a scholarly treatise on the Latin language, we again find the notion of primeval words (*verba primigenia*), but we also have something that did not exist in Plato's day: a theory of lexical morphology comprising not only word formation but also nominal and verbal inflection. In an inflectional paradigm, a vast number of word forms issue from a single lexical unit. "If from these words there are one thousand primitives," writes Varro, "then from their 'declensions' there can be five hundred thousand different forms since from any individual primitive word about five hundred forms are made by 'declensions.' " [33] Thus, we have the notion here of a large set of forms being generated from a single source—essentially something comparable in language to roots in nature.

In referring to these sources, Varro uses a number of biological metaphors, for example, the words *nasci* 'to be born,' *radix* 'root,' and *stirps* 'trunk,' 'root or stock from which a family springs,' and he lists a number of primordial verbs, which he says have their own roots, and contrasts them with inflected verbs (*verba declinata*), which come from (*oriuntur*) other verbs. [34] In another passage, he uses the term "root" to refer to the etymological source of a Latin word in another language; of the word *subulo* 'flute player' he says, "Its roots are to be sought in Etruria not in Latium." [35]

But other terms he uses in this connection have no unequivocal biological connotations, such as *origo* 'origin' itself and the corresponding verb *oriri* 'to arise,' 'originate,' which were used to refer to both biological and nonbiological phenomena, as a glance at any reasonably comprehensive Latin lexicon reveals. Similarly, *declinare* has no biological connotations; like the term *casus*, it suggests the image of deviating from the true direction, "a recta flecti, curvari, deflectere, divertere" (Egidio Forcellini, *Lexicon totius Latinitatis*). *Derivare*, another mainstay of the etymologist's terminology, meant originally "to drain or convey water from its regular course, derive or turn off into a different channel" (Forcellini), and hence when it was used in grammatical contexts it was an analogy drawn from the terminology of irrigation.

In the Latin grammars of late antiquity, two of which (the works of Donatus and Priscian) survived as popular textbooks throughout the Middle Ages and into modern times, word composition was emphasized at the expense of derivation. Derivative formatives (prefixes, suffixes, etc.) were not isolated, nor were inflectional endings cited in isolation. Composition was regarded as resulting from concatenating words, and the changes which words undergo in that process were mentioned, but there was no notion of the stem. Thus, in Donatus' *Ars minor* (mid-fourth century A.D.), word compounds were classified depending on which of the components appeared unchanged, there being four possibilities: both components could be unchanged, both could undergo change, or either the first or the second component could appear distorted. [36] The word *corruptus* is the term Donatus uses to refer to the changed member. Many derivatives in the modern sense he considers to be simplex, for example, *docte* 'learnedly' and *prudenter* 'prudently.'

At the same time, participles (the participle was considered a separate part of speech in traditional grammar) were said to "come from" verbs: two participles were said to come from active verbs, three from deponents, and so forth.[37] Moreover, in the *Ars maior,* Donatus says that adverbs are either derived from other parts of speech or original, and he uses the term *nasci* 'to be born.'[38]

Inflection was handled in terms of base forms and paradigms. The base form for each noun was the nominative singular, and the base form for a verb was the first person singular present indicative active. The grammarians of late antiquity, however, did not always leave the paradigms to speak for themselves (as is done in some traditional grammars of modern times), but often derived the various members of a paradigm from one another by simple logical operations. Thus, Priscian derives the genitive plural form of first-declension nouns (e.g., *poetarum* 'poets') from the ablative singular (*poeta* 'poet') by the addition of the syllable *rum*: "ut 'ab hoc poeta, horum poetarum.'"[39] The accusative singular of the same declension (e.g., *poetam*) was said to be formed from the genitive-dative form (*poetae*) by replacement: "Accusativus *ae* diphthongum genetivi sive dativi mutat in *am,* ut 'poetae, hunc poetam.'"[40] The nominative singular form was never derived, even in instances where this might have been an advantageous way to handle the facts. Thus, Priscian derives the genitive singular *sapientis* from the nominative singular *sapiens* 'wise' by removing the final *s* and adding *tis*.[41] Here many modern linguists would find it more convenient to posit a stem *sapient-* underlying all the inflected forms and then to derive the genitive by adding the ending *-is* to that stem, and the nominative by adding *-s* to the stem, with a general rule to reduce the resulting consonant cluster *nts* to *ns*.

Augustine, who was writing a short time after Donatus, is of special interest in that as a Christian writer he incorporates Jewish linguistic speculations stimulated by the Old Testament into his view of the origin of language. For example, in the *City of God* we have in chapters 4 and following of book 16, a discussion of the Tower of Babel story, and in that connection a clear statement of the view that Hebrew is the oldest language of mankind, that primeval linguistic unity was ended by the direct intervention of God against the builders of the Tower of Babel, that mankind was from then on linguistically disunited, that these new scattered linguistic fragments gave rise to ethnic divisions (not the other way round!), and that the biblical account of the genealogy of the sons of Noah in Genesis provides an enumeration of the tribes and nations of the contemporary world.

Augustine regarded Hebrew as the universal language spoken before Babel, which was retained by Heber, the eponymous founder of the Hebrews:

Similarly, when because of impious pride the nations were punished and divided by the diversity of languages and the city of the impious received the name of confusion, that is to say Babylon, there was one family, that of Heber, in which what had previously been the language of all was destined to remain. . . . Since, therefore, this language survived in his family at the same time as the other nations were divided by different languages (this language having been previously common to all mankind, as is justifiably believed), the language was subsequently called Hebrew.[42]

This view that Hebrew was the original speech of mankind seems to have been derived from Jewish tradition,[43] perhaps from a comment in the Jerusalem Targum (Gen. 11:1), a work that the Babylonian Talmud ascribes to Onkelos the proselyte. If that is true, the notion may go back as far as the second century A.D. Presumably, it was transmitted to Augustine by way of Jerome, who was active in the middle to late fourth century and who was a student of Hebrew (and incidentally a pupil of Donatus). In his *Commentarii in Sophoniam prophetam,* Jerome refers to the belief of the Jews that Hebrew was spoken by all mankind before the building of the Tower of Babel and that it would again be the universal language after the coming of the Messiah.[44] In one of his letters, Jerome expresses his own view regarding the origin of language: "The whole of antiquity has handed down the belief that the Hebrew language, in which the Old Testament is written, is the beginning of speech and universal language, and the beginning of all the languages that we speak."[45]

The subsequent wide dissemination of this notion throughout the Latin Middle Ages was ensured by its mention by Isidore of Seville (c. A.D. 602–636) near the beginning of his popular *Etymologiae.* The important points here, from our point of view, are (1) the key notion that from a single language (Hebrew) one could proceed to a multiplicity of languages, or in other words that Hebrew was the protolanguage of all human languages, and (2) the fact that Isidore calls Hebrew the "mother" of all other languages and scripts: "linguam Hebraicam omnium linguarum et litterarum esse matrem" (*Etymologiae* 1.3.4).

In another work traditionally ascribed to Augustine, the *Principia dialecticae,* we find an exposition of Stoic etymological theory. The Stoics, Augustine reports, derive words from other words and those words in turn from other words, until they reach a point where the things being referred to and the sounds of the words have some similarity. As Augustine expresses it: "Consider a little by what means they [the Stoics] imagine that one arrives at that ultimate cradle of words, beyond which they do not believe that one should look for further origins on the grounds that were somebody to attempt to discover any he would find none."[46]

The ultimate roots with which the Stoics operated turn out to be words of contemporary Latin, for example, *hinnitus* 'neighing,' *balatus* 'bleating,' *clangor* 'blare (of trumpets).' This is because, apart from occasional lexical items transmitted in legal and religious apothegms, the etymologist working in antiquity had little if any knowledge of earlier stages of his language. In practice, therefore, he was engaged in a hopeless enterprise. Lacking knowledge of the history of Greek or Latin, he was committed to a theory that could only have been convincingly demonstrated by means of evidence that was not available to him. Hence, the echoic or synesthetic roots were conveniently assumed to have survived intact. It is curious to observe, however, that in spite of the questionable character of their basic assumptions, etymological theorists did not hide behind modesty. As Augustine puts it, "The Stoics *assert* that no word exists for which a definite explanation cannot be found."[47]

In the face of criticism from skeptics (including the author of the *Principia dialecticae* himself), this approach to etymology persisted throughout antiquity, the Middle Ages, and the early modern period—indeed, until the advent of comparative philology at the beginning of the nineteenth century. Thus, Julius Caesar Scaliger, in an influential book published in the mid-sixteenth century, argues:

Although etymology is in many instances obscure, and although different people propose different explanations for the same word, it is by no means to be abandoned, for it needs to be investigated all the more because it is so mysterious. What is more hidden than the truth? But in many things, truth is the most important thing to be desired, a fact which nobody would be so shameless as to deny.[48]

Thus, etymology is once more put on a par with truth: if investigating truth is hazardous, the study of etymology is no less so. To cast doubts on etymology would then be tantamount to expressing doubts about the possibility of knowledge in general.

Indeed, the patent absurdity of many etymologies proposed by various authors, both ancient and modern, was at times regarded as giving etymologists unlimited license to unleash their speculative imagination, as in a bizarre confession once made by Jacques Dubois, a sixteenth-century grammarian of French, who addresses his readers as follows: "You should not be surprised if I offer some rather unreasonable etymologies (several of them will strike you as such), since you can find much more unreasonable ones in Probus, Marcellus, Varro, Perotti, Calepino, and other Latin etymologists, not to speak of Suidas, Hesychius, the *Etymologicum,* and others."[49]

Summing up the evidence we have so far examined, we must conclude that the seeds of biologism had been sown in the area of etymology long before the advent of comparative linguistics. It is surely significant that in his influential *Poetics* (first published in 1561), the elder Scaliger defined etymology as the process of "generating" one word from another.[50]

Let us now turn to the analytical system developed by grammarians of the Semitic languages and its impact on Westerners in the sixteenth century. In the case of Hebrew, the discovery of the seven conjugations (the so-called *bin-yanim* 'formations'), is attributed to Moses Qimḥi, who lived in Spain in the twelfth century. The use in Hebrew grammatical analysis of the terms "base" (*yesodh*) and "root" (*shoresh*) is apparently older, the former being ascribed to Gaon Saadya (d. 942) and the latter to Judah Ḥayyūj (c. A.D. 1000). In Arabic grammar, a number of terms corresponded to the Hebrew *yesodh* and *shoresh*, namely *jaðr* 'root,' *asl* 'origin,' and *masdar* 'source,' 'spring,' 'fountain.' The Islamic grammatical tradition was already flourishing in the seventh century of the Christian era, and the notion of the root was utilized by grammarians of Arabic before the Hebrew *yesodh* and *shoresh* came into use as technical grammatical terms among Jews.[51]

In the Hebrew grammatical tradition, the *shoresh* 'root' was the third person singular perfect *qal* form, for example, *qaṭal* 'he killed.'[52] Typically, the *shoresh* was found to consist of three consonants, in the case just cited *q-ṭ-l*. The fundamental meaning of a lexical item was thought to depend on these three radical consonants, and accessory notions were said to be indicated by changes in the vowels occurring between the consonants and by affixes occurring before or after the root. Moreover, the root was assumed to be basic not only to the verbal forms but also to nominal derivatives. Consequently, in Hebrew dictionaries nouns were listed not separately but under the corresponding verbal roots, namely, the *qal* form of the verb in question.

Another difference between Hebrew and Western grammatical practice was that the Hebrew grammarians isolated affixes or formatives, that is, prefixes and suffixes. Thus, derived from *qaṭal* 'he killed' we have *tiqṭolnah* 'they kill,' containing the prefix *ti-* and the suffix *-nah*. In this connection it is interesting to note that the English words "affix," "prefix," and "suffix" are terms that have been introduced only relatively recently. Neither the terms nor the underlying concepts occur in the grammars of Latin written in antiquity and the Middle Ages. The *Oxford English Dictionary* reports the word "suffix," for instance, no earlier than 1614.

Given their way of handling morphology, grammarians of Hebrew organized dictionaries not by words but by roots. To look up a word the dictionary user had to know from what root it was derived. This system is by no means an artifact of the grammatical imagination. Hebrew words are on the whole easy to decompose into root and formative elements. In fact, one might say that because of the triconsonantal framework of so many of the words, Hebrew as a language is etymologically transparent. Moreover, Hebrew script, if we ignore the so-called *matres lectionis,* provided letter symbols for the consonants only. Therefore, grammarians customarily classified letters into two types: radical letters (consonants that occur in roots) and "servile" letters (the symbols occurring in prefixes and suffixes).

Finally, Hebrew grammarians, like their Islamic predecessors and models, were more sophisticated in their approach to the articulatory relations among speech sounds than Western grammarians were. Thus, the phonetic affinity of homorganic stops and fricatives was revealed by the use in the Masoretic script of the point traditionally referred to by the term *dagesh lene* (the symbols for the stops *b, d,* and *g* contain the dagesh, while the symbols for the corresponding homorganic fricatives lack it). Moreover, Hebrew grammarians classified the consonants of the language according to place of articulation, and they even had phonetic terms to refer to the resulting classes. In contrast, traditional Western grammarians before the sixteenth century seldom if ever referred to the points of articulation and had no clear conception of the manners of articulation (e.g., stop versus fricative).

A number of features of Hebrew grammatical practice were therefore novel to Western scholars in the late Renaissance. One of the first Hebrew grammars to appear in the West, the *De rudimentis Hebraicis,* by the German humanist Johannes Reuchlin (1455–1522), published in Pforzheim in 1506, was modeled closely on the system of David Qimḥi.

It is fascinating to observe Reuchlin imparting this new way of treating language to his readers. For this purpose, he quotes a concocted Latin derivative (*inhonorificabilitudines*) and progressively divests it of all its affixes until only the bare root (*honor*) is left, concluding with the words "and now we apply the same procedure in Hebrew."[53] It is interesting that he expounds the theory of the triconsonantal root without resorting to the word "root" itself. Instead, he uses the traditional Western term *primitivum,* and the image he suggests is that of removing clothes from a body until it is naked. But note the procedure he follows in stripping off the various affixes one by one until the root is laid bare. To my knowledge, this is the first time that a Western grammarian actually broke words down in a step-by-step fashion.

It was not long before grammatical theorists and methodologists in the West began to recommend that these ideas be applied to the study of language in general. The most explicit suggestion I have seen along these lines occurs in a book with the suggestive title *De ratione communi omnium linguarum et literarum,* written by a sixteenth-century Swiss Hebraist, Theodor Bibliander (d. 1564), who was professor of Old Testament in Zurich for many years.[54]

The purpose of Bibliander's book was to demonstrate that all languages, both civilized and uncivilized, can be described grammatically and that this can and should be done in a uniform manner. As the author says at the beginning of the first chapter:

My purpose in writing this treatise has been to expound a systematic method whereby not only written languages which have been cultivated by outstanding minds but also the uncivilized and barbaric languages of all nations could be embraced, and whereby these languages could be explained and comprehended in an appropriate order and with greater accuracy, and become better adapted for all written and spoken uses. Such a method will make it possible to inspect all languages and evaluate them more precisely and compare them with each other, either in part or as wholes.[55]

Two interesting features of Bibliander's book are his ideas on language relationship and his attempt to show that the Hebrew system of listing roots instead of words in dictionaries should be used in displaying the origins of words in all languages. With regard to the latter, he says: "In my opinion, the arrangement observed hitherto in describing the Hebrew language, which involves listing the roots (*origines*) in alphabetical order and subsuming derivatives and compounds under the appropriate root *like offspring belonging to the same lineage,* is suitable for investigating all languages."[56] He also recommends listing formatives in the manner of the Hebrew grammarians: "Every language has formative and characteristic letters and syllables, and if they are arranged in order, as is done in grammars of Hebrew, this contributes much to a deeper knowledge of languages."[57] He asserts that all scholars agree in regarding the Hebrew language as "principem et parentem" (*De ratione communi,* p. 4). On the relationship of Hebrew and Arabic, he says: "That Arabic (or Punic) is descended from the Hebrew language is proved by the letters, words, and inflectional analogy."[58]

Bibliander also raises the issue of whether Ethiopic is related to Hebrew, Arabic, and Aramaic, and in that connection he quotes a passage from a book by a contemporary, the French orientalist Guillaume Postel (1510–1581), which reads as follows:

It [Ethiopic] is so closely related to Hebrew, Chaldean, and Arabic that there are very few words in it which are not found in one of them. Although, however, they differ very often in vowel points, the verb themes or roots either originate from the Hebrew ones, or are related to the Chaldean, or are similar to Arabic words, due to geographic propinquity and social intercourse.[59]

It is clear that grammatical affinity is a crucial consideration here. Thus, Postel, as quoted by Bibliander, has the following to say about Ethiopic: "Although the shapes of the letters and the left-to-right direction in which they are written differ from Hebrew, nevertheless the phonetic values of the letters, the origins of words, and the inflectional patterns are very similar."[60]

It is also clear that what Bibliander had in mind was a total comparison of the oriental languages with Western languages, not merely as regards word origins but also in morphology and syntax. After pointing out that both Jewish and non-Jewish grammarians of Hebrew had described the language in terms of definite rules (*certis legibus*), he says:

And one will be at liberty to organize Greek, Latin, German, or any other language in the same way and to compare them with Hebrew as regards letters, syllables, and the meanings, origins, inflection, compounding, and syntax of words. Accordingly, these languages when compared with Hebrew will have been described in the same system and in terms of a common theory (*doctrina*). The best theory will be the one which agrees most closely with reason and nature.[61]

We are therefore dealing with the notion that some languages are related to each other, with the obvious implication that other languages are not related in that fashion. This latter point is clarified by a statement made by Conrad Gesner, a compatriot and contemporary of Bibliander. Gesner asserts in his *Mithridates,* published in Zurich in 1555, that some languages are related to one another in varying degrees (he uses the word *cognatae*), whereas others are completely unrelated (*omnino distant*): "It would, however, appear to be not so much meddlesome curiosity as a pursuit worthy of an educated man to know which languages are more or less related to one another and which are completely unrelated."[62]

In Gesner's book, moreover, we note the occasional use of the term "mother" to refer to a parent community. In discussing the Hungarians, for example, the author quotes information provided in Aeneas Sylvius Piccolomini's *Asia* (mid-fifteenth century; Piccolomini died as Pius II in 1464) to the effect that there was another ethnic group of the same name much farther east which was the "parent" of the European Hungarians or, as Gesner himself puts it, "Not far from the source of the river Don, there is another Hungary, the *mother* of the one we have been discussing, quite similar in language and customs, although ours is more civilized and worships Christ, while the other one lives after the manner of barbarians and worships idols."[63] I point out, however, that Piccolomini himself had not used the word *mater* in the passage in question in his *Asia*.

It seems likely, therefore, that the use of the term "mother language", in the sense of a parent language, derives naturally from this usage of *mater* in an ethnological context and also undoubtedly from Isidore's use of *mater* as the primeval parent of all existing languages. An attempt to enumerate all the language families known at the time was then simply a question of time. Joseph Justus Scaliger (1540–1609) wrote a short essay entitled "Diatriba de Europaeorum linguis," which appeared in a collection of his writings published in Paris a year after his death. Using a genealogical image, Scaliger defines mother languages (*linguae matrices*) as languages that differ from one another completely and that give rise to dialects or descendants (*propagines*):

We may term languages from which many dialects are derived like descendants "mother languages." The descendants of a single mother language are linked to each other by some degree of intercourse. There is no relationship among mother languages, either in words or in analogy. . . . Therefore, let our mother languages be those which differ from one another in all respects. Eleven such mother languages, and no more, remain in the whole of Europe.[64]

Using another biological image, he speaks of mother languages as spawning descendants (he uses the verb *pario* 'to bear, bring forth'). Thus, of the Romance languages he says, "The DEUS mother language [viz., Latin] brought forth Italian, French, and Spanish."[65] Scaliger also makes it clear that a word in the mother language appears in various divergent forms in the descendant languages:

Identical words cause a language to look like a single language, while modifications of these same words yield each descendant. Thus, we call Italian, Spanish, and French Latin because the same Latin word appears changed in the three of them. For example, Latin *gener* ['daughter's husband'] is *genero* for the Italians, *yerno* for the Spanish, and *gendre* for the French. They are Latin words if you consider their origin, but if you consider their differences each of the three languages can claim the word as its own.[66]

As regards the use of "root" in the sense of a base form in morphological analysis, there are a number of examples from the early modern period of use of the word itself in the description of individual languages. First, the term *radix* and its German counterpart *Wurzel* were already part of the grammatical terminology of German by the end of the sixteenth century, as witnessed by the *Teutsch Grammatick oder Sprach-Kunst* by Laurentius Albertus (1573) and Sebastian Helber's *Teutsches Syllabierbüchlein* of 1593.[67] Further afield, in the Japanese grammar of the Spanish Jesuit Diego Collado, published in 1632 by the Congregation for the Propagation of the Faith, the notion of the root is invoked in connection with the verb: "On their own, verb roots do not indicate tense. Therefore, to designate tense verbs must be formed from them and conjugated."[68] Note that Collado says "on their own"; he might have added "unlike the Hebrew root." It is possible, in other words, that Collado had the Hebrew root in mind here. The term "root" then figures prominently throughout the subsequent description of the Japanese verbal inflection. On the other hand, Collado does not handle the noun in this way, presumably because it is uninflected, case relations in Japanese being indicated by post-positions, "particles which are placed after the nouns and indicate case differences in all nouns."[69]

At the same time, the use of the notion of base forms from which other inflectionally related forms could be generated was apparently fairly widespread. Thus, in the grammar of Quechua (an important South American Indian language) by Domingo de Santo Tomás, which appeared in Valladolid in the middle of the sixteenth century, the author uses the first person singular indicative of each verb as the base form (*fundamento*) from which to derive all the other forms: "La primera persona del singular del indicativo no se forma de otra alguna porque ella es el fundamento de la formacion delas demas, y ellas se forman della."[70] Here we have the concept of the root without the term itself.

The view of language relationship among etymologists in the seventeenth and eighteenth centuries was a further elaboration of the ideas we have seen expressed by Joseph Scaliger.[71] Thus J. G. Wachter, in his *Glossarium Germanicum* (1737), pictures linguistic diversity beginning in human families and eventually giving rise to dialects, which in turn spawn separate languages, which give rise to further dialects, and so on ad infinitum. He also emphasizes that this whole process is gradual and continuous, since nature does not make jumps (*natura non facit saltus*), a familiar Leibnizian dictum. He explicitly asserts that Western languages are derived from oriental languages and emphasizes that all existing languages are derived from earlier ones—in other words, there is no such thing as an "unrooted" language.[72]

The situation just before the advent of comparative linguistics is mirrored in the introductory chapter of Johann Christoph Adelung's monumental *Mithridates* (1806). In Adelung's view, language was originally "hörbarer Ausdruck des Empfundenen" (p. xi). He argues eloquently for the echoic origin of the primordial roots; he speaks of what he terms "Nachbildung des Gehörten" (ibid.). The number of primordial roots for all the languages of the world is, according to him, quite small:

If one adds up the various roots of all languages they are found to amount to no more than a few hundred; they recur in all languages from the Ganges to the St. Lawrence, and in each language they must suffice to express the incalculably large number of concepts which the human race in all regions of the world and at all levels of culture not only possesses but is capable of acquiring in all future ages. . . . In short, the number of roots does not amount to more than a few hundred in any language. All roots recur in every language, and the totality of this small stock of roots constitutes the basis of all present and future linguistic wealth.[73]

That there was a link between the notion of comparative grammar and the genealogical classification of languages can be clearly seen in a book review written by August Wilhelm Schlegel published in 1803. The book being reviewed by Schlegel was August Ferdinand Bernhardi's *Sprachlehre* (Berlin, 1801–1803), Bernhardi being a follower of the great classical scholar Friedrich August Wolf. Schlegel maintains that a description of the rules of usage peculiar to some particular language (*specielle Grammatik*, as against *allgemeine Grammatik*) remains a mass of lifeless information until the investigator comprehends the organism of the language and can understand the individual peculiarities of the language "historically." In Schlegel's view, characterizing the individuality of each language is a task that grammarians have so far not achieved. What is needed to remedy this, Schlegel says, is a comparative grammar designed to bring out the features peculiar to each language as well as the features shared by different languages. Thus, Greek and Latin could be so compared, then the Germanic and the Romance languages (with English as intermediate between them), and then finally all the previously mentioned languages could be compared with each other as members of a single language family and their similarities and differences enumerated. The same procedure could be applied to the oriental (i.e., Semitic) languages, and they could finally be compared collectively with the Western languages.[74]

It is interesting to note the extent to which comparative grammar, as conceived by Schlegel, was still intimately connected with typological considerations. It is also clear that, at this point, Schlegel still has a place for universal grammar of the kind represented by Bernhardi's book. The compatibility of universal grammar and comparative grammar, in the period immediately before the appearance of Boppian linguistics, is also obvious from an interesting remark made by the French orientalist Silvestre de Sacy in his *Principes de grammaire générale* (1799). De Sacy regards the physiological description of speech sounds as the basis of both "etymological science" and "comparative grammar," and it is clear that these are things of which he does not disapprove.[75]

Franz Bopp was interested in the origin of grammatical forms, which means in practice that he restricted etymological speculation to what we now call affixes, that is, all morphological elements other than roots. That he did indeed think in terms of such a restriction is explicitly stated in the preface to the first edition of his *Vergleichende Grammatik*: "The only thing that we leave untouched is the mystery of the roots, i.e., the rationale for the names of the primordial concepts. Thus, we do not investigate why, for instance, the root *ī* means 'go' and not 'stand,' or why the sound sequence *sthā* or *stā* means 'stand' and not 'go.'"[76]

Bopp attributed to the primordial period of grammatical formation the same perfection of linguistic insight that earlier etymologists had attributed to the primeval period during which the assignment of words to concepts or things had supposedly first taken place. The original system of inflections reconstructed by Bopp and ascribed by him to the prehistoric period was, he believed, gradually abandoned as people lost the vividness of their linguistic intuition. Thus, he portrays the progressive abandonment of the dual number in most Indo-European languages in the following terms: "The dual, like the neuter gender, is the first to disappear in course of time as the vividness of sensual conception is weakened, or its use progressively withers away, and it is then replaced by the abstract plural which encompasses indefinite plurality."[77]

Thus, much of Bopp's approach to comparative grammar constitutes a retention, unconscious perhaps, of ideas or attitudes characteristic of earlier linguistic speculation. First, he seems to have believed that in approaching a parent language such as Indo-European the linguist is dealing with a period of linguistic development close to the origin of language itself. In other words, Indo-European for Bopp was a primitive language in a qualitative not merely a historical sense.

Second, Bopp tended to believe that while roots were inherited in each of the extant Indo-European languages the process of compounding and agglutination took place subsequently and to a large extent independently in the various daughter languages. One might say, therefore, that his parent language was characterized not by *prōta onomata* 'primordial words' but by *prōtai rhizai* 'primordial roots.'

Third, the period since the parent language stage is for him a period of decay and degeneration, as it had been in the analogous Stoic and medieval theories. Moreover, the two most salient manifestations of linguistic degeneration were the change in phonetic rules (Bopp's own term was *Wohllautsgesetze* 'euphonic laws'), in other words, the process that linguists were later to call sound change (*Lautwandel*), and the tendency to indulge in more and more compounding and agglutination of roots.

Fourth, we note that Bopp's parent language is endowed with rationality and even with self-comprehension, features which he assumes were progressively lost in the subsequent period of decay. One recalls that the Stoics had held similar views with regard to the character of language in its formative stages.

Fifth, Bopp was as interested in the problem of classifying languages typologically as his immediate predecessors had been. The Indo-European languages are characterized by him as having monosyllabic roots, unlike the Semitic languages with their triconsonantal roots. One recalls that Adelung had organized his entire survey of the world's languages on the basis of the number of root syllables: monosyllabic languages, polysyllabic languages, and so on. In its beginning stages, comparative grammar was intimately connected with traditional etymology on the one hand, and with the typological characterization of languages on the other, as our quotations from de Sacy and Schlegel have shown.

Finally, Bopp was a strong supporter of the idea that proof of linguistic relationship should be based on a comparison of inflectional systems. As we have seen, this principle had already been formulated in the sixteenth century by scholars whose ideas on language relationship were influenced by their acquaintance with Semitic languages. Hence, the affixal material on which Bopp focused his attention performed much the same function as the *verba primigenia* had done for the etymologists of antiquity.

I say all this not to belittle Bopp but to point out that many of the conceptual ingredients of early nineteenth-century linguistics had been inherited from the past. This amounts to a perfectly commonplace situation in intellectual history and in no way detracts from Bopp's achievement. One might say that the basic conceptual picture in Bopp's mind was one that had taken over two millennia to take shape. In brief, it had the following ingredients: the notion of a primeval language giving rise to descendant languages, the notion of a primordial lexicon and a principle of grammatical formation inherited from that primeval language and active in all its descendants, and finally a process of phonetic decay operating on initially transparent morphological formations.

The first of these notions had for the most part biblical origins—it was the linguistic analogue of the genealogy of the descendants of Noah. The second was rooted in Greek antiquity—the notion of primordial words (*prōta onomata*). The principle of grammatical formation was a grammatical analogue to that same notion of primordial roots. Finally, the notion of phonetic decay must also be attributed to the ancient Greeks, whose speculations are reflected in Plato's *Cratylus*.

So we are dealing here with concepts derived from various facets of the intellectual climate of the past two and a half millennia, not with direct applications to the study of language of notions developed by natural historians. Linguistic genealogy is a logical corollary of the notion that the diversity of languages is linked to the diversity of ethnic groups, which in turn was thought to be a function of descent from the sons of Noah, the sole survivor of the Flood. This is not to rule out the possibility that some measure of cross-fertilization of the two fields of biology and linguistics may have occurred in the middle of the nineteenth century, but genealogical thinking had been deeply entrenched in the study of language for a long time. This was recognized by Schleicher in his *Darwinsche Theorie:*

First, as regards Darwin's assertion that species change in course of time, a process repeated time and again which results in one form arising from another, this same process has long been generally assumed for linguistic organisms. . . . We set up family trees of languages known to us in precisely the same way as Darwin has attempted to do for plant and animal species.[78]

Indeed, even earlier in the century, the great Danish linguist Rasmus Rask had admitted on one occasion that when he was composing his famous prize essay of 1818 on the genetic affiliations of Icelandic he lacked a clearly worked out taxonomic system of the kind that Linnaeus had introduced in botany.[79] Furthermore, it is important to recall that linguists had always regarded the notion that languages change as axiomatic, while the idea of the mutability of biological species was not widely accepted until well into the nineteenth century.

In other words, there is no clear evidence that developments in linguistics up to the beginning of the nineteenth century had been influenced in any essential way by natural history. That is, although as historians we can now see striking parallels between the two fields, and although the early practitioners of comparative grammar at times showed signs of being aware of them, it would clearly be a mistake to interpret them as evidence of a causal relation.

As for the use of tree diagrams, this occurred independently in linguistics and in natural history. Lamarck, for instance, has a genealogical tree in his *Philosophie zoologique* of 1809.[80] Traditionally, the relation between biological species had been seen as an infinitely graded one-dimensional *scala naturae,* which logically excluded representation in the form of a family tree. But in the eighteenth century signs that indicate a degree of dissatisfaction with this picture began to appear. Thus, in the middle of that century, Vitaliano Donati proposed that "natural progressions," as he put it, "should be compared to a network rather than a chain."[81] To my knowledge, the first person to suggest the tree as a logically conceivable way of representing the relations among animal species was the German naturalist Peter Simon Pallas in a book published in 1766.[82] As regards linguistic works, we have a branching diagram of the so-called Thracian language group in an unpublished essay written by Rasmus Rask in 1819.[83] Tree diagrams seem to have been first used by linguists, therefore, after the notion of evolutionary transformation had already taken hold.

In biology, the situation is the reverse. While Lamarck had certainly come to espouse evolutionary views by the time he published his *Philosophie zoologique* in 1809, it would perhaps strain credibility to portray Pallas as an evolutionist. After Lamarck but before Darwin, evolutionary ideas were again defended by Robert Chambers in a much-criticized book published anonymously in 1844.[84] Chambers called his theory "the doctrine of progressive development." On the other hand, branching diagrams of the Porphyrian variety occur in botanical treatises as early as the seventeenth century. The earliest I am aware of is in a book by the English naturalist Robert Morison (1620–1683) which appeared in Oxford in 1680.[85] Morison's example was quickly followed by John Ray in his *Methodus plantarum nova*[86] and by Rivinus in a work that appeared in 1690.[87]

The use of Porphyrian trees is logically independent of the notion of historical evolution, but when a field shifted to an evolutionary perspective, as natural history did in the course of the nineteenth century, such diagrams performed a useful function in representing the genetic relations between whatever entities the discipline dealt with. I would like to suggest that they came to be widely used in both biology and linguistics in the nineteenth century because the two logical ingredients, namely, hierarchical taxonomy and the concept of evolutionary development, existed in both, though in language study the latter had made its appearance a great deal earlier than in natural history. But the question of the rise of the evolutionary perspective is extremely complex and deserves more study.[88] We may some day find solid documentary evidence that the Linnaean system influenced the comparative study of languages in the eighteenth century, or that the comparative anatomists at the end of that century stimulated linguists into applying similar techniques to theirs in the study of language. Meanwhile, it is perhaps well to hold the ultimate answer to such questions in suspense.

Notes

1. "Nous sommes . . . profondément convaincus que quiconque pose le pied sur le terrain de la *langue* peut se dire qu'il est abandonné par toutes les analogies du ciel et de la terre" (Ferdinand de Saussure, *Cours de linguistique générale: Édition critique par Rudolf Engler* 1 [Wiesbaden: Otto Harrassowitz, 1968], p. 169). Compare Saussure's caustic remarks on the illusions (he calls them "mirages") that language has created in the popular mind (ed. cit., p. 23).

2. John Lyons, *Introduction to Theoretical Linguistics* (Cambridge: Cambridge University Press, 1968), p. 9. Henri Fleisch, in his study of Islamic philology, is even more outspoken. The Arabic grammatical tradition, he says, "n'est pas fondé en réalité, est une violence faite à la langue, n'a pas de valeur scientifique, est caduc et n'a plus qu'un intérêt historique" (*Traité de philologie arabe* 1: *Préliminaires, phonétique, morphologie nominale,* Recherches publiées sous la direction de l'Institut de lettres orientales de Beyrouth 16 [Beirut: Imprimerie Catholique, 1961], p. 16). One may also compare Alexander Sperber's caustic comments on the tradition of Hebrew grammar in his *Historical Grammar of Biblical Hebrew* (Leiden: A. J. Brill, 1966), p. 15. General linguists have written a great deal on the relation between traditional Western grammar and contemporary academic linguistics. Before the advent of generative grammar, they were for the most part decidedly hostile to traditional grammar; see, e.g., Leonard Bloomfield, *Language* (New York: Holt, 1933), p. 3. More recently attitudes have changed, largely under Chomsky's influence (see Noam Chomsky, *Aspects of the Theory of Syntax* [Cambridge, Mass.:

M.I.T. Press, 1965], pp. 63–64), but there are still dissenting voices (see, e.g., Roy Harris, *The Language Myth* [New York: St. Martin's Press, 1981]).

    3. Laurenz Lersch, *Die Sprachphilosophie der Alten,* 3 vols. (Bonn: H. B. Konig, 1838–1841; reprint, Hildesheim and New York: Georg Olms Verlag, 1971).

    4. H. Steinthal, *Geschichte der Sprachwissenschaft bei den Griechen und Römern,* 2 vols., 2d ed. (Berlin: F. Dümmlers Verlag, 1890–1891; reprint, Bonn: F. Dümmlers Verlag, 1961).

    5. Paolo Rotta, *La filosofia del linguaggio nella patristica e nella scolastica* (Turin: Fratelli Bocca, 1909), esp. pp. 14–29.

    6. Ernst Cassirer, *Philosophie der symbolischen Formen* 1 (Berlin: Bruno Cassirer Verlag, 1923), esp. pp. 106–112.

    7. Arno Borst, *Der Turmbau von Babel: Geschichte der Meinungen über Ursprung und Vielfalt der Sprachen und Völker,* 4 vols. in 6 (Stuttgart: Anton Hiersemann, 1957–1963).

    8. James Stam, *Inquiries into the Origin of Language* (New York: Harper & Row, 1976).

    9. See esp. Jan Agrell, *Studier i den äldre språkjämförelsens allmänna och svenska historia fram till 1827,* Uppsala Universitets Årsskrift 1955, 13 (Uppsala: Lundequistska Bokhandeln and Wiesbaden: Otto Harrassowitz, 1955); R. H. Robins, "The History of Language Classification," in *Current Trends in Linguistics* 11: *Diachronic, Areal, and Typological Linguistics,* ed. Thomas A. Sebeok (The Hague and Paris: Mouton, 1973), pp. 3–41; Henry M. Hoenigswald, "On the History of the Comparative Method," *Anthropological Linguistics* 5, no. 1 (January 1963): 1–11. On the important notion of the family tree, see Henry M. Hoenigswald, "Schleicher's Tree and Its Trunk," *Ut videam: Contributions to an Understanding of Linguistics, for Pieter Verburg on the Occasion of His 70th Birthday* (Lisse: Peter de Ridder Press, 1975), pp. 157–160.

    10. A concise statement of the way linguists view the relationship of languages and the relationship of individual linguistic features may be found in Alan S. C. Ross, "Philological Probability Problems," *Journal of the Royal Statistical Society,* Series B (Methodological), vol. 12 (1950): 19–59, esp. pp. 19–21, and see the ensuing discussion of Ross' paper (pp. 41–59, esp. pp. 41, 49). Linguists in the English-speaking world now prefer the terms "protolanguage," "Proto-Germanic," "Proto-IndoEuropean," etc., to "Primitive Germanic," "Primitive Indo-European," etc. It is interesting to note that the prefix "proto-" came into currency in linguistics and biology at about the same time (viz., in the final decades of the nineteenth century), if the entries in the *Oxford English Dictionary* can be relied upon.

    11. See Antoine Meillet, *La méthode comparative en linguistique historique,* Instituttet for sammenlignende kulturforskning, Publikationer, ser. A: Forelaesninger, 2 (Oslo: H. Aschenhoug & Co.; Cambridge, Mass.: Harvard University Press, 1925), pp. 15–17.

    12. Philologists were talking about genealogical diagrams as early as the first half of the eighteenth century. Thus, in an edition of the Greek New Testament that appeared in 1734, the editor, Johann Albrecht Bengel, wrote: "Posset variarum lectionum ortus, per singulos codices, per paria codicum, per syzygias minores maioresque, per familias, tribus, nationesque illorum, investigari et repraesentari, et inde propinquitates discessionesque codicum ad schematismos quosdam reduci, atque ita res tota per *tabulam quandam quasi genealogicam* oculis subici, ad quam tabulam quaelibet varietas insignior cum agmine suorum codicum, ad convincendos etiam tardissimos dubitatores, exigeretur. Magnam coniectanea nostra silvam habent; sed manum de tabula, ne risuum periculo exponatur veritas" (*Hē kainē diathēkē: Novum Testamentum Graecum* [Tübingen, 1734], p. 387 (emphasis added), quoted in Caspar René Gregory, *Textkritik des Neuen Testaments* 2 [Leipzig: J. C. Hinrichs, 1902], p. 908). For further discussion of these issues, see Giorgio Pasquali, *Storia della tradizione e critica del testo* (Florence: Felice Le Monnier, 1952), p. 9, and Sebastiano Timpanaro, *La genesi del metodo del Lachmann,* 2d ed., Biblioteca di Cultura (Padua: Liviana Editrice, 1981), pp. 24ff. Note that Timpanaro points out the link be-

tween Lachmann and Jacob Grimm (*La genesi*, p. 83 n. 3) and the affinity between textual criticism and comparative grammar (pp. 45–47, 81ff.).

13. See August Schleicher, *Compendium der vergleichenden Grammatik der indogermanischen Sprachen*, 4th ed. (Weimar: Hermann Böhlau, 1876), p. 9; and *Die Darwinsche Theorie und die Sprachwissenschaft*, 2d ed. (Weimar: Hermann Böhlau, 1873), after p. 33.

14. Schleicher's original statement reads: "Die länge der linien deutet die zeitdauer an, die entfernung der selben von einander den verwantschaftsgrad" (*Compendium*, 4th ed., p. 8). In this respect, Schleicher believed that his family trees differed from Darwin's, which he claimed presented merely an idealized schema: "wobei man nicht ausser Acht lasse, dass Darwin ein ideales Schema aufstellt, wir aber das Bild der Entstehung einer gegebenen Sippe zeichnen" (*Die Darwinsche Theorie und die Sprachwissenschaft*, 2d ed. (Weimar: Herman Böhlau, 1873), pp. 15–16.

15. See *Scholia in Aristotelem*, ed. Christianus Augustus Brandis, Aristotelis opera 4, edidit Academia Regia Borussica (Berlin: G. Reiner, 1836), p. 34, and the article "Porphyrian" in the *Oxford English Dictionary*. The Lullian "tree of knowledge" is also a purely schematic method of representation, although Lull's trees, unlike Porphyry's, look superficially like real ones; see Ramón Lull, *Arbor scientiae* (Barcelona: Pere Posa, 1485; Lyons: Gilbert de Villiers, 1515). Similar to Lull's schema is a type of tree diagram to be found in Iohannes de Pastrana, *Grammatica Pastrane* (Lisbon: Pedro Bonhomini de Cremona, 1512); for facsimiles, see Manuel II, *Livros antigos portuguezes, 1489–1600* 1 (London: Maggs Bros., 1929), pp. 229–230. An early use of genealogical trees in the form of roundels connected by lines occurs in the *Compendium historiae in genealogia Christi,* usually attributed to Peter of Poitiers, first chancellor of the University of Paris (d. 1205); see *La grande encyclopédie* 26 (Paris: Société anonyme de la Grande encyclopédie, n.d.), p. 900.

16. To my knowledge, the first linguist to raise his voice publicly in opposition to the *Stammbaum* picture was Hugo Schuchardt (1842–1927) in his *Probevorlesung* at the University of Leipzig in 1870, subsequently published as a monograph entitled *Über die Klassifikation der romanischen Mundarten* (Graz: K. K. Universitätsbuchdruckerei "Styria," 1900); see *Hugo Schuchardt-Brevier: Ein Vademecum der allgemeinen Sprachwissenschaft*, ed. Leo Spitzer, 2d ed. (Halle: Max Niemeyer, 1928), pp. 166–188. Earlier still, there are clear adumbrations of his position in Schuchardt's monograph *Der Vokalismus des Vulgärlateins* 1 (Leipzig: B. G. Teubner, 1866), pp. 82–83. Similar ideas were then expressed by Johannes Schmidt in a much-cited monograph entitled *Die Verwantschaftsverhältnisse der indogermanischen Sprachen* (Weimar: Hermann Böhlau, 1872) and by the influential theorist Hermann Paul; see his *Principien der sprachgeschichte* (Halle: Max Niemeyer, 1880), chap. 12, pp. 231ff.; *Prinzipien der Sprachgeschichte*, 4th ed. (Halle: Max Niemeyer, 1909), chap. 2, pp. 37ff.

17. See Max Hermann Jellinek, *Geschichte der neuhochdeutschen Grammatik von den Anfängen bis auf Adelung* 2 (Heidelberg: Carl Winter, 1914), p. 133.

18. In a grammar of Hebrew widely used in the seventeenth century, "theme" is defined thus: "Thema sive radix apud Hebraeos ex tribus communiter constat literis, rarissime ex pluribus" (Johann Buxtorf, *Epitome grammaticae Hebraeae* [Cambridge: Roger Daniel, 1646], sig. A5r). Buxtorf, in fact, uses *thema* more frequently than *radix*. This is perhaps an indication that "root" was still a relatively unfamiliar term at that time and that grammarians preferred the traditional "theme."

19. See his *Vergleichende Grammatik* 1 (Berlin: Druckerei der Königl. Akademie der Wissenschaften, 1836), pp. 133ff.

20. See Wilhelm Gesenius, *Hebrew Grammar,* 14th ed., rev. by E. Rödiger (New York: D. Appleton, 1852), p. 19.

21. "Die Flexionen machen den wahren Organismus einer Sprache aus" (*Vergleichende Grammatik* 1, part 2, p. vii).

22. "Die halb schon ausserhalb der grammatik liegende syntax" (*Über den Ursprung der Sprache* [Berlin: F. Dümmler's Verlag, 1852], p. 7).

23. See, e.g., Arthur O. Lovejoy, *The Great Chain of Being: A Study of the History of an Idea* (Cambridge, Mass.: Harvard University Press, 1936), pp. 227ff. As regards Friedrich Schlegel, the genealogical tree is already mentioned in his seminal 1808 monograph on the language and wisdom of India: "Die alten Sprachen, deren Stammbaum wir von der Wurzel bis zu den Hauptästen im ersten Buche zu verfolgen suchten, sind eine Urkunde der Menschengeschichte, lehrreicher und zuverlässiger, als alle Denkmale in Stein, deren halbverfallne Riesengrösse die späte Nachwelt, zu Persepolis, Illoure, oder an dem aegyptischen Theba mit Erstaunen betrachtet" (Friedrich Schlegel, *Ueber Sprache und Weisheit der Indier: Ein Beitrag zur Begründung der Alterthumskunde* [Heidelberg: Mohr und Zimmer, 1808], p. 157).

24. "Die Sprachen sind Naturorganismen, die, ohne vom Willen der Menschen bestimmbar zu sein, entstunden, nach bestimmten Gesetzen wuchsen und sich entwickelten und wiederum altern und absterben; auch ihnen ist jene Reihe von Erscheinungen eigen, die man unter dem Namen 'Leben' zu verstehen pflegt. Die Glottik, die Wissenschaft der Sprache, ist demnach eine Naturwissenschaft; ihre Methode ist im Ganzen und Allgemeinen dieselbe, wie die der übrigen Naturwissenschaften. So konnte mir denn auch das Studium des Darwinschen Buches, zu dem Du mich veranlasstest, nicht als meinem Fache allzu ferne liegend erscheinen" (*Die Darwinsche Theorie und die Sprachwissenschaft*, 2d ed. [Weimar: Hermann Böhlau, 1873], p. 7).

25. See William Chomsky, *Hebrew: The Eternal Language* (Philadelphia: Jewish Publication Society of America, 1957), p. 118. It is thought that the first clear statement of the relationship between Hebrew and Arabic was made by Yehuda Ibn Quraysh of Tahort in North Africa, a Jewish physician who is usually assumed to have flourished at the end of the ninth century and the beginning of the tenth. He claims that the two languages differ only in a limited number of regular phonetic changes, and states, "The cause of this resemblance and (consequent) interchange is to be found in the propinquity of habitations and consanguinity of races" (Hartwig Hirschfeld, *Literary History of Hebrew Grammarians and Lexicographers* [London: Oxford University Press, Humphrey Milford, 1926], p. 18). Compare David Tene, "The Earliest Comparisons of Hebrew with Aramaic and Arabic," *Progress in Linguistic Historiography*, ed. Konrad Koerner, Amsterdam Studies in the Theory and History of Linguistic Science, Series 3, 20 (Amsterdam: John Benjamins, 1980), pp. 355–377. On the early history of Hebrew grammatical writing, see Wilhelm Bacher, *Die Anfänge der hebräischen Grammatik, und Die hebräische Sprachwissenschaft vom 10. bis zum 16. Jahrhundert*, Amsterdam Studies in the Theory and History of Linguistic Science, Series 3: Studies in the History of Linguistics 3 (Amsterdam: John Benjamins, 1974); W. Jacques van Bekkum, "The 'Risāla' of Yehuda Ibn Quraysh and Its Place in Hebrew Linguistics," *Historiographia Linguistica* 8 (1981): 307–327. A useful general survey of the history of Hebrew linguistics is provided by the article "Hebrew Linguistic Literature," *Encyclopaedia Judaica* 16 (Jerusalem: Keter, 1971), cols. 1352–1401.

26. See Lersch, *Sprachphilosophie der Alten*, 3: 3–7.

27. The secondary literature on the *Cratylus* is vast. See Charles H. Kahn, "Language and Ontology in the *Cratylus*," *Exegesis and Argument: Studies in Greek Philosophy Presented to Gregory Vlastos*, ed. E. N. Lee, A. P. D. Mourelatos, and R. M. Rorty, *Phronesis*, Supplementary Volume 1, 1973 (New York: Humanities Press, 1973), pp. 152–176; Rotta, *Filosofia del linguaggio*, pp. 14–29; Steinthal, *Geschichte der Sprachwissenschaft*, 1: 79–113; Antonino Pagliaro, "Struttura e pensiero del 'Cratilo' di Platone," *Nuovi saggi di critica semantica* (Messina and Florence: G. D'Anna, 1956), pp. 49–76; Josef Derbolav, *Der Dialog "Kratylos" im Rahmen der platonischen Sprach- und Erkenntnisphilosophie* (Saarbrücken: West-Ost Verlag); Jetske C. Rijlaarsdam, *Platon über die Sprache: Ein Kommentar zum Kratylos* (Utrecht: Bohn,

Scheltema & Holkema, 1978). An excellent French translation by Louis Méridier is provided with a copious introduction; see Platon, *Oeuvres complètes* 5, pt. 2: *Cratyle,* texte établi et traduit par Louis Méridier, Collection des Universités de France (Paris: Société d'édition "Les belles lettres," 1931), pp. 7–48.

28. Thus, as late as the early nineteenth century, in Immanuel Bekker's edition of the *Cratylus,* the rival positions of Hermogenes and Cratylus are summarily stated, during the course of which the following assertion is made concerning Socrates' opinion: "Socrates vero, cuius vero illi [scil. Cratylus et Hermogenes] se iudicio subiciunt, in Cratyli sententiam magis inclinare videtur" (*Platonis . . . scripta Graece omnia . . . diligenter enotavit Immanuel Bekker* 4 [London: excudebat A. J. Valpy, A. M. sumptibus Ricardi Priestley, 1826], p. 186).

29. For a statement of the modern approach to etymology, see Joh. Nikolai Madvig's illuminating article "Einige Voraussetzungen der Etymologie und ihre Aufgabe," in his *Kleine philologische Schriften* (Leipzig: B. G. Teubner, 1875), pp. 319–355. The traditional view of etymology, on the other hand, is well expressed in a textbook of logic from the seventeenth century: "Etymologia est ratio vocis in significando, ex origine declarata; veluti, cum *mutuum* dicitur, quasi *de meo tuum*" (Franco Burgersdijck, *Institutionum logicarum synopsis* [Cambridge: John Field, 1666], p. 18). This is very close to the definition given by Isidore of Seville a thousand years earlier: "Etymologia est origo vocabulorum, cum vis verbi vel nominis per interpretationem colligitur" (*Etymologiae* 1.29.1). The term "etymology" therefore subsumed both synchronic morphology and what we today call etymology, i.e., the earlier history of word forms. But this usage was not considered vague and confusing because until the second half of the nineteenth century no distinction was drawn between synchronic rules of formation and the history of forms. It is perhaps the generation of linguists born in the 1840s, with their emphasis on correct chronology, who were responsible for disentangling these two notions.

30. Varro, *De lingua Latina* 7.11. For more etymologies of this kind, see Johannes von Arnim, *Stoicorum veterum fragmenta* 2 (Leipzig: B. G. Teubner, 1902), p. 47.

31. On *testamentum,* see Book VI, chap. 36, *Laurentii Vallae opera* (Basle: Henrichus Petrus, 1543), sig. O5r, and compare Aulus Gellius, *Noctes Atticae,* 6.12.

32. On Varro and Stoic etymology, see Hans Joachim Mette, *Parateresis: Untersuchungen zur Sprachtheorie des Krates von Pergamon* (Halle and Saale: Max Niemeyer, 1952); Karl Barwick, *Probleme der stoischen Sprachlehre und Rhetorik,* Abhandlungen der sächsischen Akademie der Wissenschaften zu Leipzig, Philologisch-historische Klasse, Band 49, Heft 3 (Berlin: Akademie-Verlag, 1957); and Hellfried Dahlmann, *Varro und die hellenistische Sprachtheorie,* 2d ed. (Berlin and Zurich: Weidmannsche Verlagsbuchhandlung, 1964).

33. "Horum verborum si primigenia sunt ad mille . . . ex eorum declinationibus verborum discrimina quingenta milia esse possunt ideo, quod a singulis verbis primigeniis circiter quingentae species declinationibus fiunt" (6.36). By *declinationes* Varro is referring to inflected forms as well as derivatives. Thus, as examples he quotes various inflected forms of the verb *lego* 'I read' (*leges, lege, lecturus*) together with a number of derivatives: *lectio* 'the act of reading,' *lector* 'reader.'

34. "Primigenia dicuntur verba ut *lego, scribo, sto, sedeo* et cetera, quae non sunt ab alio quo verbo, sed suas habent radices. Contra verba declinata sunt quae ab alio quo oriuntur, ut ab *lego legis, legit, legam,* et sic indidem hinc permulta" (6.37).

35. "Radices eius in Etruria, non Latio quaerundae" (7.35).

36. "Quibus modis nomina componuntur? Quattuor: ex duobus integris, ut *suburbanus;* ex duobus corruptis, ut *efficax, municeps;* ex integro et corrupto, ut *insulsus;* ex corrupto et integro, ut *nugigerulus;* aliquando ex compluribus, ut *inexpugnabilis, imperterritus*" (Heinrich Keil, ed., *Grammatici Latini* 4 [Leipzig: B. G. Teubner, 1864], p. 355, ll. 21–25).

37. Keil 4.363–364.
38. "Adverbia aut a se nascuntur, ut *heri, hodie, nuper,* aut ab aliis partibus orationis veniunt" (Keil 4.385.12ff.).
39. Heinrich Keil, *Grammatici Latini* 2 (Leipzig: B. G. Teubner, 1855), p. 292, l. 1.
40. Keil 2.285.13–14.
41. Keil 2.281.10–11.
42. "Ita, quando merito elatioris impietatis gentes linguarum diversitate punitae atque divisae sunt, et civitas impiorum confusionis nomen accepit, hoc est appellata est Babylon, non defuit domus Heber, ubi ea quae antea fuit omnium lingua remaneret. . . . Quia ergo in eius familia remansit haec lingua, divisis per alias linguas ceteris gentibus, quae lingua prius humano generi non immerito creditur fuisse communis, ideo deinceps Hebraea est nuncupata" (*De civitate Dei* 16.11).
43. See Edward Ullendorff, "The Knowledge of Languages in the Old Testament," *Bulletin of the John Rylands Library* 44 (1961–1962): 455–465, esp. 455; reprinted in Edward Ullendorff, *Is Biblical Hebrew a Language? Studies in Semitic Languages and Civilizations* (Wiesbaden: Otto Harrassowitz, 1977), pp. 37–47, esp. p. 37.
44. Referring to a passage in the author he is commenting on, Jerome says: "Haec Iudaei interpretantur in adventu Christi, quem sperant venturum esse, et dicunt universis gentibus congregati et effuso super eas furore Domini in igne zeli eius terram devorandam. Et sicut ante aedificationem turris fuit, quando una lingua omnes populi loquebantur, ita conversis omnibus ad cultum veri Dei, locuturos Hebraice et totum orbem Domine serviturum" (J. P. Migne, *Patrologia Latina* 25 [Paris: Garnier Frères, 1884], col. 1378).
45. "Initium oris et communis eloquii et hoc omne quod loquimur Hebraeam linguam, qua vetus testamentum scriptum est, universa antiquitas tradidit" (*Sancti Eusebii Hieronymi Epistulae,* I: *Epistulae I–LXX,* ed. I. Hilberg, Corpus scriptorum ecclesiasticorum Latinorum 54 [Vienna: F. Tempsky; Leipzig: G. Freytag, 1910], p. 82).
46. J. P. Migne, *Patrologia Latina* 32 (Paris: Garnier Frères, 1877), col. 1413. The text as it appears in Migne's edition reads as follows: "Vide tamen paululum quomodo perveniri putant ad illa verborum cunabula, vel ad stirpem potius atque adeo sementum, ultra quod quaeri originem vetant, nec si quis velit, potest quidquam invenire." The phrase "ad stirpem potius atque adeo *sementum* [*sic*]" is bizarre, but exactly the same wording occurs in Rudolf Schmidt, *Stoicorum grammatica* (Halle: Eduard Anton, 1839), p. 26. Schmidt used a Paris edition of 1586, whereas Migne had at his disposal a Basel edition printed in 1558. Emendation would seem to be called for here.
47. "Stoici autumant, quos Cicero in hac re irridet, nullum esse verbum cuius non certa ratio explicari possit" (Migne, *Patrologia Latina* 32, col. 1412); emphasis added.
48. "Etymologia vero etsi in multis obscura est, superque eadem voce alia alii visa, tantum tamen abest ut tollenda sit, ut tam maxime sit investiganda quam maxime latet. Quid enim occultius veritate? At multis in rebus ea in primis desideratur, neque tamen quisquam tam sit impudens qui eam neget" (Julius Caesar Scaliger, *De causis linguae Latinae libri tredecim* [Lyons: apud Seb. Gryphium, 1540], sig. x3r).
49. "Nec te mirari oportet quod etyma quaedam absurdiuscula (qualia tibi forte videbuntur nonnulla) tradidimus, cum multo absurdiora apud Probum, Marcellum, Varronem, Perottum, Calepinum, et alios Latinorum etymographos invenias, ut interim Suidam, Hesychium, Etymologicum, ceterosque taceam" (Iacobus Sylvius Ambianus, *In linguam Gallicam isagoge, una cum eiusdem grammatica Latino-Gallica, ex Hebraeis, Graecis, et Latinis authoribus* [Paris: Robertus Stephanus, 1531], sig. a5v).
50. "Etymologia est vocis ratio, id est vis qua vox a voce generatur" (*Poetices libri septem,* Book 2, chap. 89 [Heidelberg: In Bibliopolio Commeliniano, 1607], sig. x3v).

51. See Wilhelm Bacher, *Die Anfänge der hebräischen Grammatik, und Die hebräische Sprachwissenschaft vom 10. bis zum 16. Jahrhundert*, Amsterdam Studies in the Theory and History of Linguistic Science, Series 3: Studies in the History of Linguistics 3 (Amsterdam: John Benjamins, 1974), pp. 136–137; "Hebrew Grammar," *The Jewish Encyclopedia* 6 (New York and London: Funk & Wagnalls, 1904), pp. 67–80, esp. 68–71; Fleisch, *Traité de philologie arabe* 1, p. 3 (see n. 2, above); Hirschfeld, *Literary History of Hebrew Grammarians and Lexicographers*, p. 7 (see n. 25, above); R. H. Robins, *A Short History of Linguistics*, 2d ed., Longman Linguistics Library 6 (London and New York: Longman, 1979), p. 98; W. Keith Percival, "The Reception of Hebrew in Sixteenth-Century Europe: The Impact of the Cabbala," *Historiographia Linguistica* 11 (1984): 21–36; Gérard Troupeau, "La notion de 'racine' chez les grammairiens arabes anciens," Sylvain Auroux et al., eds., *Matériaux pour une histoire des théories linguistiques* (Lille: Université de Lille, 1984), pp. 239–246.

52. See *Gesenius' Hebrew Grammar*, ed. E. Kautzsch and A. E. Cowley, 2d Eng. ed. (Oxford: Clarendon Press, 1910), pp. 114ff.

53. "Ex his omnibus per nos hactenus expositis facilius scies id quod in lingua Hebraica maxime dicitur artis esse ac ingenii singularis laudatissimum indicium, cum omnis dictio aut primitiva sit aut derivativa, exuere nomen derivativum singulis vestibus ad nuditatem quousque appareat primitivum. Latine prius tecum agam. Exempli gratia, si hoc nomen pluralis numeri *hae inhonorificabilitudines* velis ad primitivum statum redigere, quid facies? Decortabis illud omnibus tunicis quousque absolutum appendiciis appareat primitivum, sic: Reici primo articulum *hae*. Secundo, compositionem primam qua ligatur cum praepositione *in*. Tertio, compositionem qua ligatur cum *ifica*, a *facio*. Quarto, terminationem verbalem aptitudinis, quae est *bilis*. Quinto, terminationem abstractionis, quae est *tudo*. Sexto, casualem inflectionem genitivi, quae est *din*. Septimo, numeralem terminationem *es*. Separatis *hae, in, ifica, bili, tudin, es*, quid remanet? Certe nudum illud, quod nominamus *honor*. Ecce primitivum. Nunc pariter operare in Hebraicis, ad cuius faciliorem aditum scire te iubeo. Quod apud Hebraicis secundum communem et vulgarem morem plerumque omnia primitiva tres habent literas, quibus suum cuique corpus constituitur" (*De rudimentis Hebraicis* [Pforzheim: Th. Anshelm, 1506], pp. 582–583). It goes without saying that a form such as *inhonorificabilitudines* is a grammarian's invention. It may be compared with the word *honorificabilitudinitas*, quoted in the so-called *Petri grammatici excerpta*; see *Grammatici Latini, Supplementum: Anecdota Helvetica quae ad grammaticam Latinam spectant collecta edidit Hermannus Hagen*, ed. Heinrich Keil (Leipzig: B. G. Teubner, 1870), p. 164, l. 17. I am indebted to Professor Brian Ó'Cuív of the Dublin Institute for Advanced Studies for bringing this passage to my attention.

54. *De ratione communi omnium linguarum et literarum commentarius Theodori Bibliandri* (Zurich: Froschoverus, 1548). On Bibliander, see George J. Metcalf, "Theodor Bibliander (1504–1564) and the Languages of Japheth's Progeny," *Historiographia Linguistica* 7 (1980): 323–333.

55. "Proposui . . . hoc commentario . . . scribere methodum et rationem qua non modo literatae illae et a praeclaris ingeniis excultae linguae verum etiam rudes et barbarae linguae omnium nationum comprehendi possint totae, et apto quidem ordine necnon rectius explicari et percipi facilius et accommodari melius ad omnem usum scribendi et dicendi, qua denique ratione linguae omnes perspici et iudicari exactius queant et conferri mutuum vel tote vel in qualibet parte" (*De ratione communi*, p. 1).

56. "Censeo enim ordinem servatum hactenus in Ebraicae linguae explanatione, ut origines litterarum ordine digerantur, et cuilibet origini subiciantur ab eo derivata et composita et copulata, ceu propagines stirpi suae, aptum esse omnibus linguis cognoscendis" (ibid., p. 169; emphasis added).

57. "Esse in omni lingua litteras et syllabas formativas et characteristicas, quas ordine dis-

ponere, ut fit in Ebraicis grammaticis, multum conducere ad penitiorem cognitionem linguarum" (ibid., p. 163).

58. "Arabicam seu Punicam linguam propaginem esse Ebraici sermonis arguunt litterae et dictiones et analogia inflectionis" (ibid., p. 4).

59. "Est adeo Hebraicae, Chaldicae et Arabicae affinis, ut rara sit admodum dictio quae non reperiatur in aliarum aliqua. Quamvis autem saepissime punctis differant, tamen verborum themata seu radices frequentissime sunt aut Hebraeis ob originem, aut Chaldeis ob affinitatem, aut Arabibus ob viciniam et convictum assimiles" (ibid., p. 6).

60. "Etsi vero litterarum formae et ordo scribendi a laeva dextrorsum discrepet a lingua Ebraica, tamen litterarum potestas, origines dictionum, et inflectendi formulae proxime accedunt" (ibid., p. 5).

61. "Et licebit eodem ordine vel Graecam linguam vel Latinam vel Germanicam vel aliam quamcumque linguam comprehendere et conferre cum Ebraea in literis, in syllabis, et verborum significatione, origine, inflectione, compositione, structura. Itaque linguae istae universae quae componentur ad hunc modum cum Ebraea, sub eandem rationem et doctrinam communem redactae sunt. Illa vero erit optima doctrina et ars quae cum ratione et natura maxime quadrabit" (ibid., p. 4).

62. "Videtur autem non tam curiosa quam liberalis cognitio, ut quae inter se cognatae sint linguae plus minus, quae omnino distent, intelligamus" (Conrad Gesner, *Mithridates: De differentiis linguarum tum veterum tum quae hodie apud diversas nationes in toto orbe terrarum in usu sunt . . . observationes* [Zurich: Froschoverus, 1555], sig. A2r).

63. "Extat adhuc non longe ob ortu Tanais altera Hungaria, nostrae huius de qua sermo est *mater*, lingua et moribus paene similis, quamvis nostra civilior est, Christi cultrix, illa ritu barbarico vivens servit idolis" (Gesner, *Mithridates,* sig. G4v; emphasis added).

64. "Linguas matrices vocare possumus ex quibus multae dialecti tamquam propagines deductae sunt. Propagines quidem unius matricis linguae commercio inter se aliquo coniunctae sunt; matricum vero inter se nulla cognatio est, neque in verbis neque in analogia. . . . Sunto igitur nobis matrices eae quae per omnia inter se discrepant, cuiusmodi undecim, non amplius, supersunt in universa Europa . . ." (Joseph Justus Scaliger, *Opuscula varia antehac non edita* [Paris: apud Hadrianum Beys, 1610], p. 119). See Hans Arens, *Sprachwissenschaft: Der Gang ihrer Entwicklung von der Antike bis zur Gegenwart,* Orbis academicus I/6, 2d ed. (Freiburg and Munich: Karl Alber, 1969), pp. 74–76; and R. H. Robins, *Short History of Linguistics,* pp. 165–167.

65. "Matrix DEUS peperit Italicam, Gallicam, et Hispanicam . . ." (*Opuscula varia,* p. 121).

66. "Eadem verba faciunt unam linguam videri, sed eorundem verborum traiectio, immutatio, inflexio aliam atque aliam propaginem facit. Nam Italicam, Hispanicam, et Gallicam Latinam vocamus propter unum verbum Latinum, quamquam varie immutatum in illis tribus; exempli gratia *gener* Latinum Italis est *genero,* Hispanis *yerno,* Gallis *gendre.* Latina sunt si originem spectes, sin distinctionem una quaeque natio harum trium illud vindicat sibi" (ibid., p. 119).

67. See Jellinek, *Geschichte der neuhochdeutschen Grammatik* 2:135 (n. 17, above).

68. "Radices verborum de se non dicunt tempus; unde ut illud dicant debent formari verba et coniugari" (*Ars grammaticae Iaponicae linguae . . . composita . . . a Fr. Didaco Collado Ordinis Praedicatorum* [Rome: Typis et impensis Sac. Congr. de Propag. Fide, 1632], p. 18).

69. "Particulae quae postpositae nominibus casuum differentias constituunt in omnibus nominibus" (ibid., p. 6).

70. Raúl Porras Barrenechea, ed., *Grammatica o arte de la lengua general de los Indios de los reynos del Perú por el maestro Fray Domingo de Santo Tomás,* Universidad Nacional

Mayor de San Marcos, Publicaciones del cuarto centenario (Lima: Edición del Instituto de histo-ria, 1951), p. 79. This is a facsimile of the edition printed in Valladolid between 1560 and 1561. A similar statement is made in the lexicon by the same author, which followed closely on the footsteps of the grammar, see *Lexicon o vocabulario de la lengua general del Peru,* ed. Raúl Porras Barrenechea (Lima: Edición del Instituto de historia, 1951), p. 12.

71. See George J. Metcalf, "The Indo-European Hypothesis in the Sixteenth and Seven-teenth Centuries," in *Studies in the History of Linguistics: Traditions and Paradigms,* ed. Dell Hymes, Indiana University Studies in the History and Theory of Linguistics (Bloomington: In-diana University Press, 1974), pp. 233–257.

72. "Sunt qui linguam nostram ex Campis Sinear immediate progressam et post dis-persionem humani generis statim in his terris auditam arbitrantur, quos ego non refutandos sed explodendos existimo. Nam natura non facit saltum, neque in locorum intervallis, neque in tem-pore, neque in ulla alia re, sed ex proximo tendit ad propinqua, a propinquis ad remotiora, et a remotioribus non nisi longo temporis tractu ad remotissima progreditur. Itaque ex una et primi-tiva lingua (quod sacrae litterae et gentium consensus testantur) suscitatae sunt primo variae dia-lecti et totidem paene variationes quot hominum familiae at familiarum cognationes, quibus deinde separatis dialecti paulatim abierunt in linguas. Ex his linguis postea formatae sunt novae linguae et linguarum dialecti et dialectorum linguae, non subito et repente sed gradatim et pede-temptim, temporum, locorum, migrationum, et coloniarum adiumento. Hunc enim natura con-tinuo servavit ordinem et quasi legem in novis sermonibus producendis, ut recentiores fierent ex antiquioribus, occidentales ex orientalibus. Ex vi huius legis impossibile erat ullam in toto ter-rarum orbe exsistere linguam quae non in aliqua priore fuerit radicata" (Johann Georg Wachter, *Glossarium Germanicum* [Leipzig: J. F. Gleditsch, 1737], sig. a3ra).

73. "Wenn man die verschiedenen Wurzellaute aller Sprachen zusammen zählet, so finden sich deren nicht mehr als wenige hundert, welche in allen Sprachen von dem Ganges bis zum Lorenz-Flusse wieder kommen, und in jeder hinreichen müssen, die unermessliche Menge Be-griffe, welche das menschliche Geschlecht unter allen Zonen, und in allen Graden der Cultur nicht allein hat, sondern auch in allen künftigen Weltaltern noch erwerben kann, auszudrücken. . . . Genug, die Zahl der Wurzellaute beträgt in keiner Sprache über wenige hundert; jeder dieser Laute kommt in jeder Sprache von neuem vor, und dieser kleine Vorrath zusammen genommen macht den Grund des ganzen gegenwärtigen und künftigen Sprachreichthums aus" (Johann Christoph Adelung, *Mithridates, oder allgemeine Sprachenkunde,* Erster Theil [Berlin: In der Vossischen Buchhandlung, 1806], pp. xvi–xvii). The notion that roots were originally few in number and have not been added to since, absurd as it may seem to us today, was still held in the middle of the nineteenth century by the practitioners of comparative grammar—witness the fol-lowing remark in Max Müller's *Lectures on the Science of Language:* "If you consider that, what-ever view we take of the origin and dispersion of language, nothing new has ever been added to the substance of language, that all its changes have been changes of form, that no new root or radical has ever been invented by later generations, as little as one single element has ever been added to the material world in which we live; if you bear in mind that in one sense, and in a very just sense, we may be said to handle the very words which issued from the mouth of the son of God, when he gave names to 'all the cattle, and to the fowl of the air, and to every beast of the field,' you will see, I believe, that the science of language has claims on your attention, such as few sciences can rival or excel" (F. Max Müller, *Lectures on the Science of Language* 1, 6th ed. (London: Longmans, Green, 1880), pp. 28–29.

74. "An die allgemeine Sprachlehre kann sich die specielle Grammatik für einzelne Sprachen mit grossem Vortheile anschliessen. Ehe jene nach philosophischen Principien aufgestellt ist, bleibt für diese der Sprachgebrauch eine todte Gedächtnissache. Ist man hingegen über den gesetzmässigen Organismus der Sprache überhaupt im Klaren, so können die hinzukommenden

besondern Bestimmungen als das Individuelle historisch begriffen und charakterisiert werden. Bei den Meistern des Stils ist das Gefühl für die Individualität ihrer Sprache sehr rege, allein von Grammatikern ist bis jetzt für die Charakteristik wenig geleistet worden. Die vergleichende Grammatik, eine Zusammenstellung der Sprachen nach ihren gemeinschaftlichen und unterscheidenden Zügen, würde dazu ungemein behülflich sein. So müsste man das Griechische und Lateinische; die Sprachen deutschen Stammes, das Deutsche, Dänische, Schwedische und Holländische; die neolateinischen mit deutschen und andern Einmischungen; das Provenzalische, Französische, Italiänische, Spanische, Portugiesische; dann das in der Mitte liegende Englische; endlich wieder alle zusammen als eine gemeinschaftliche Sprachfamilie nach grammatischen Uebereinstimmungen und Abweichungen und deren innerem Zusammenhange vergleichen. Eben so die orientalischen erst unter sich, hernach mit den occidentalischen. Leichter ist es zwar diesen Plan zu entwerfen, als ihn auszuführen; doch würde solchergestalt die Philologie immer mehr zur Kunst werden, und auch die Ausbildung der lebenden Sprachen kunstmässiger fortschreiten können" (August Wilhelm Schlegel, *Sämmtliche Werke* 12, ed. Eduard Böcking [Leipzig: Weidmann'sche Buchhandlung, 1847], pp. 152–153).

    75. "Les considérations . . . sur les organes naturels de la parole, et sur la nature et la variété des sons et des articulations que nous produisons avec le secours de ces organes, sont sans doute bien propres à nous faire admirer la puissance et la sagesse du Créateur, et la fécondité des moyens dont il se sert pour parvenir à ses fins. Elles ont aussi une grande utilité pour les recherches qui appartiennent à la science étymologique et à la Grammaire comparative des idiomes anciens ou modernes. Mais elles n'appartiennent pas à la première instruction, pas même à l'instruction commune et usuelle. C'est la nature et l'imitation qui enseignent aux enfans l'usage qu'ils doivent faire de leurs organes, pour la prononciation de la langue maternelle; et c'est encore de l'imitation, et non de la connoissance des organes et des principes de la Grammaire, que l'homme plus avancé apprend à imiter des articulations étrangères au langage du pays où il a pris naissance" (A. I. Silvestre de Sacy, *Principes de grammaire générale* [Paris: A. A. Lottin, 1799], pp. iii–iv).

    76. "Nur das Geheimniss der Wurzeln oder des Benennungsgrundes der Urbegriffe lassen wir unangetastet; wir untersuchen nicht, warum z. B. die Wurzel *I* gehen und nicht stehen, oder warum die Laut-Gruppirung *STHA* oder *STA* stehen und nicht gehen bedeute" (*Vergleichende Grammatik* 1: iii).

    77. "Der Dual geht wie das Neutrum im Laufe der Zeit mit der Schwächung der Lebendigkeit sinnlicher Auffassung am ersten verloren oder wird in seinem Gebrauch immer mehr verkümmert, und dann durch den abstrakten, die unendliche Vielheit umfassenden Plural ersetzt" (ibid., 1: 136).

    78. "Was nun zunächst die von Darwin behauptete Veränderungsfähigkeit der Arten im Verlaufe der Zeit betrifft, durch welche . . . aus einer Form mehrere Formen hervorgehen (ein Prozess der sich natürlich abermals und abermals wiederholt), so ist sie für die sprachlichen Organismen *längst allgemein angenommen* . . . Von Sprachsippen, die uns genau bekannt sind, stellen wir eben so Stammbäume auf, wie diess Darwin . . . für die Arten von Pflanzen und Thieren versucht hat" (*Die Darwinsche Theorie und die Sprachwissenschaft*, 2d ed., pp. 14–15; emphasis added). It is also important to recall that Schleicher had read and been influenced by other naturalists before reading Darwin's *Origin of Species*. For instance, in the opening contribution to the journal *Beiträge zur vergleichenden Sprachforschung* in 1858, he starts out by invoking a principle enunciated in Matthias Jacob Schleiden's *Grundzüge der wissenschaftlichen Botanik:* "Wenn Schleiden in seinen grundzügen der wissenschaftlichen botanik den grundsatz ausspricht und durchführt 'dass die einzige möglichkeit zu wissenschaflicher einsicht in der botanik zu gelangen . . . das studium der entwickelungsgeschichte sei,' so müssen wir diesen für das reich der organischen naturwesen überhaupt geltenden satz auch für die sprachwissenschaft schon

aus dem grunde gelten lassen, weil auch die sprachen natürliche organismen sind und die sprachwissenschaft einen theil der naturgeschichte des menschen bildet" ("Kurzer abriss der geschichte
der slawischen sprachen," *Beiträge zur vergleichenden Sprachforschung* 1 [1858], 1–27, quotation on p. 1).

79. "Tror jeg ikke jeg har erindret at berette Hr. Professoren mit ny System til Indeling af
alle Sprogene paa Jorden, hvilket jeg anser for ligesaa nødvendigt i denne Videnskab (Lingvistik)
som det linaeiske i Botaniken. . . . Da jeg skrev min Afhandling var slet intet bestemt Klassifika-
tionssystem indfaldet mig, Adelungs i Mitridat efter de fem Verdensdele duer ligesaalidt som det
gamle botaniske efter de fire Aarstider" (*Breve fra og til Rasmus Rask* 1, ed., Louis Hjelmslev
[Copenhagen: Ejnar Munksgaard, 1941]), p. 382. Similarly, in his monograph on Rask, Paul Diderichsen lists a number of startling parallels between Rask's system of taxonomy and Cuvier's,
but he admits that there is no positive proof that Rask had actually ever read Cuvier's *Leçons
d'anatomie comparée;* see Paul Diderichsen, *Rasmus Rask og den grammatiske tradition*, Det
Kongelige Danske Videnskabernes Selskab, historisk-filosofiske meddelelser 38: 2 (Copenhagen:
Ejnar Munksgaard, 1960), p. 23, cf. 234.

80. *Philosophie zoologique, ou exposition des considérations relatives à l'histoire naturelle des animaux* 2 (Paris: Dentu [et] l'auteur, 1809), p. 463, and compare vol. 1 of the same
work, pp. 277–280. John C. Greene has reproduced this tree diagram from Lamarck's *Philosophie zoologique* in his *Death of Adam: Evolution and Its Impact on Western Thought* (Ames,
Iowa: Iowa State University Press, 1959), p. 163. On Lamarck's view of the mutability of species,
see Henri Daudin, *Cuvier et Lamarck: Les classes zoologiques et l'idée de série animale
(1790–1830),* 2 vols. (Paris: Librairie Félix Alcan, 1926), Études d'histoire des sciences na-
turelles 2, esp. vol. 2, pp. 202–205.

81. "Ad una *rete* piuttosto, che ad una *catena* le naturali progressioni si dovrebbero
rassomigliare, essendo, per dir così, tessuta di vari fili, che tra loro hanno scambievole commu-
nicazione, correlazione, ed unione" (*Della storia naturale marina dell'Adriatico* [Venice: Fran-
cesco Storti, 1750], p. xxi; emphasis added).

82. *Elenchus zoophytorum* (The Hague: Apud Petrum van Cleef, 1766), pp. 23–24. On
Pallas' ideas in this regard, see Henri Daudin, *De Linné à Jussieu: Méthodes de la classification
et idée de série en botanique et en zoologie (1740–1790),* Études d'histoire des sciences na-
turelles 1 (Paris: Librairie Félix Alcan, [1926]), p. 165.

83. "Undersögelse om det gamle Nordiske eller Islandske Sprogs Slægtskab med de
asiatiske Tungemaal" [Investigation of the Relationship Between Old Norse or Icelandic and the
Languages of Asia]; see Rasmus Rask, *Ausgewählte Abhandlungen,* vol. 2, ed. Louis Hjelmslev
(Copenhagen: Levin & Munksgaard, 1932–1933), p. 9.

84. Robert Chambers, *Vestiges of the Natural History of Creation* (London: John Churchill, 1844).

85. Robert Morison, *Plantarum historiae universalis Oxoniensis pars secunda seu herba-
rum distributio nova, per tabulas cognationis et affinitatis ex libro naturae observata et detecta*
(Oxford: E Theatro Sheldoniano, 1680), see esp. pp. 360–361.

86. John Ray, *Methodus plantarum nova* (Amsterdam: Apud Janssonio-Waesbergios,
1682).

87. Augustus Quirinus Rivinus, *Ordo plantarum quae sunt flore irregulari monopetalo*
(Leipzig: Literis Christoph. Fleischeri, 1690), esp. sigs. A4v–B1r. The latter work appears as a
second volume accompanying Rivinus' influential *Introductio generalis in rem herbariam*
(Leipzig: Typis Christoph. Guntheri, 1690), in which the author discusses Robert Morison's "new
method" at some length; see sig. C2v.

88. For an interesting (albeit not entirely successful) attempt to discover deeply ingrained
theoretical perspectives at work in more than one discipline in the same historical period of the

kind we see here in the evolutionary approach common to natural history and the study of language, see Michel Foucault, *Les mots et les choses: Une archéologie des sciences humaines,* Bibliothèque des sciences humaines (Paris: Gallimard, 1966). The efficacy of Foucault's "archaeological" procedure has been much debated in recent years; see esp. Hubert L. Dreyfus and Paul Rabinow, *Michel Foucault: Beyond Structuralism and Hermeneutics,* 2d ed. (Chicago: University of Chicago Press, 1983), pp. 79–100.

2

The Life and Growth of Language: Metaphors in Biology and Linguistics

RULON S. WELLS

I

Philosophers of science have been interested in the influence of one science on another. Discernible influence is sometimes positive, sometimes negative. According to one popular view of the sciences, popular especially in Continental thought, sciences fall into essentially distinct groups, with the implications (1) that there are essential properties of one group which do not hold for any other group and (2) that therefore attempts, involving any such properties, to take a science of one group as a model for a science of another group cannot succeed. In the most common version of the view, there are two such groups: natural sciences and the science of man (*Geisteswissenschaft;* historical sciences; social sciences). Another common version, overlapping with the one just mentioned but distinct from it, is that there are sciences in which it is a mistake to apply quantitative or other mathematical methods. For example, the command not to use mathematical methods in linguistics is not a so-called categorical imperative but merely conditional: If you don't want to waste your time, get insignificant results, and so forth, don't use mathematical methods in linguistics. Kant (cf. Appendix A) has an unusual version of vitalism. Far from holding that the student of living (organic) nature should not try to bring it under inorganic laws, he encourages such attempts, but holds that they can never completely succeed. Another example of a view that emphasizes differences is that of Franz Boas—which is difficult to formulate, partly because Boas did not formulate it himself—that anthropological facts and methods differ widely from those of the physical sciences.

Besides these difference-emphasizing views there are views that emphasize likenesses, but a comprehensive survey of these does not concern us here. Rather, our present topic is a history of views that emphasize likenesses between living things (the objects of biology) and languages (the objects of linguistics). And as a further limitation of scope, we will stop a little after the middle of the nineteenth century, with the death of August Schleicher (1821–1868), glancing briefly at his critics J. N. Madvig, M. Bréal, and W. D. Whitney (on all of whom see Aarsleff 1982, and below, pp. 42, 46, 47, 51, and 58).

The intellectual influences discussed here can be described in a basically uniform way by this pattern: a set of facts or theories or methods of one science are taken as a model to be sought, or imitated, by a corresponding set of another science; thereby the latter set is compared with and assimilated to the former. Often, though not always or necessarily, the assimilation is expressed by those verbal tropes called metaphors. For certain purposes one may wish to distinguish metaphor in a strict sense (*sensu stricto*) from simile, from analogy, and so on, but such distinctions will in the main be unimportant here. Therefore I will use the word "metaphor" in a broad sense (*sensu lato*) that includes similes, positive comparisons (but not contrasts), analogies, models, patterns, and paradigms, as well as metaphors in stricter senses (it will not, however, include metonymy).

Metaphor in the strict sense has two main uses: to furnish a label and to emphasize a claimed similarity (besides these two cognitive uses, there is the affective use of arousing interest). An example of the first is "tree," in generative-transformational syntax, where the motive for using the word is primarily onomastic. The similarities claimed are slight: only of shape, and only of shape in two dimensions at that. The motive, articulated, would be something like this: "For convenience of reference, we have to call it something or other, and our name will be easier to remember if we call it by the same name as something it looks like." Thus the psychology of memory is taken into account, and association by similarity is appealed to.

Ivor Armstrong Richards, one of the leading students of metaphor in the strict sense, introduced (1936, p. 96) the valuable concepts of vehicle and tenor. (Ricoeur 1977, p. 336 n. 27 reminds us that the word "tenor" is drawn from a passage in Bishop Berkeley that Richards had quoted on p. 4.) One might have used in syntax the word that actually is used in genealogy, namely, pedigree. (Curiously, although the word "pedigree" is of French origin, *pied de gru* 'crane's foot,' present-day French speaks only of *un arbre généalogique*.) Taking the two words at their origins, one of them signifies something vegetative and the other something animal, but the two metaphorically used words, though they are different vehicles, have the same tenor. All that matters in a tree and in a crane's foot is the two-dimensional outline of its shape, in abstraction from such concrete properties as size, spatial orientation (does the "tree" have the trunk at the top, one side, or the bottom of the page?), color, and texture (cf. p. 51). Take any vehicle, and disregard all properties that you do not mean (intend) to use in your comparisons, and all those that are left comprise the tenor. Thus different vehicles may have the same tenor; those properties wherein they differ are disregarded.

Metaphor in the broad sense is assimilation, that is, positive comparison. Mental acts of comparison range from the completely undeliberate (sometimes carelessly called unintentional and sometimes loosely called unconscious) to the completely deliberate. An intermediate degree that we shall often encounter may be called the unreflective, or casual, use. Very often among our case studies, linguistic facts or theories or methods will be compared with something biological, for example, languages will be divided into living languages and dead languages. Most of these comparisons will prove to be more or less casual. They are poorly thought out; no clear discrimination is indicated between the intendedly relevant and the irrelevant; there is no discussion of whether other vehicles have the same tenor and whether the tenor could be expressed in some nonmetaphorical way. To mention two of my upcoming examples, most of those who speak of languages as alive could just as well have spoken of them as changing, and one very important person (Franz Bopp) who spoke of the "organism" of language meant no more than that words have parts. In explaining why these people spoke (or wrote) as they did, I have recourse to the philosophers' distinction between cause and reason, and between cognitive and emotive uses. Bopp had no reason to speak of the organism of language; he had only a cause. He did not care at all why his predecessors used the word "organism" metaphorically; he merely used it, in his own way, because it was current. For most of those who, over the centuries (mainly the seventeenth through nineteenth) have taken more or less seriously the conceit of languages as living things, there were other metaphorical vehicles with the same tenor available to them, as well as nonmetaphorical expressions, and it appears that they chose the biological metaphors more for their affective meaning than for their cognitive meaning.

My history chooses to end with the nineteenth century. This choice reflects my predilections, but it also reflects the facts themselves. The picture I paint, in outline, is this. The ancient and medieval worlds had hardly any sense of or feeling for history, nor therefore for time in a certain pregnant sense. The modern (postmedieval) world, groping its way toward articulating a sense of time and history, at first lisped and stammered. It metaphorized change in the race—the collective—as change in a quasi-individual, change in individuals being the only changes deeply understood by the ancient and medieval world as contrasted with the modern. Eventually (in the twentieth century) articulation became unnecessary. In this environment, the comparison of languages with living things flowered; more accurately, it was *part* of the environment. To switch to a distinctively modern metaphor, that comparison became part of the *Zeitgeist*.

When we turn to the nineteenth century, it becomes possible to distinguish between amateurs and professionals as regards the sciences of biology and linguistics. To apply the distinction to the sixteenth through eighteenth centuries is more difficult, or even impossible. A person may make valuable contributions and yet be an amateur (e.g., H. L. Mencken 1936, 1963). As linguists, I would class Rousseau, Adam Smith, and Herder as amateurs, and yet they wrote three of the most famous treatises on the origins of language. By a professional I do not mean someone who earns his living by his profession, but someone who is concerned in a deep and prolonged way with the field and is versed with previous work on the subject, and more so than with all or most other fields (instead of speaking of "professionals" I might as well speak of "specialists"). One of the main findings of my historical study is that serious use of biological metaphors is common among professional linguists through the early nineteenth century, but after that is confined mostly to amateurs. The meaning of this shift is that these metaphors ceased to be helpful; the more they were taken seriously, the more they were seen to pose difficulties. Nonmetaphorical ways of expressing their tenor were found.

William Dwight Whitney is a good representative of the attitude of nineteenth-century professionals. He subtitled his 1875 book *The Life and Growth of Language,* a title which I have appropriated for this essay and which, as he explained a decade later, signalizes a moderate position. We should be neither misled by metaphors nor afraid of them (cf. pp. 47 and 56 on Whitney). Here Whitney is an epigone of his teacher, the pre-Schleicherian Bopp; his intellectual Bildung was complete by the early 1850s.

What properties of organic life might make it a model for linguistic facts, theories, or methods? In answering such a question it is helpful to have a checklist of prominent properties; the answer to the question will then consist in some selection of these properties, in other words, in a discrimination between the relevant properties and the irrelevant properties. The metaproperty of being prominent and the metaproperty of being relevant are not absolute; they are relative to a perspective, a *Weltanschauung*. Prominence and relevance are in the eye of the *Weltanschauer*. Anthropology has taken note of this relativity by using the morpheme *ethno-*, for example, that lions are courageous is a fact of Western ethnozoology whether it is a fact of zoology or not. Two animals, both well known to a viewer, may be equally swift, and yet one may impress the viewer as more swift than another. I submit the following checklist of properties that in the Western world are generally deemed prominent properties of living things, to be discriminated into the relevant and the irrelevant for linguistic purposes.

1. Individuals change
   a. Each individual (living thing) changes, according to a pattern (the life cycle or life history). Individuals belong to kinds (see property 3). All but anomalous individuals of the same kind have the same pattern of change.
   b. Such change is continuous, not discrete (though it is not always uniform).
   c. Such change is autonomous (self-regulated).
   d. Such change is purposeful.
   e. In particular, an individual may mate with an individual and reproduce.
2. Wholeness
   a. A living individual is a unity, a whole, an organism.
   b. This is often, but ineptly, expressed thus: The whole is more than the sum of its parts. A more apt formula (Kant 1790, sec. 65) is: In an organism there is reciprocal dependence, and this reciprocal dependence can be nuanced in either of two ways: (i) every part exists for the sake of the whole or (ii) every part exists for every other part.
   c. The principle of compensation. In Aristotle, an animal that lacks weapons to defend itself against predators may compensate by being swifter than they are, that is, by not needing to defend itself. Or a species whose members are frail and vulnerable may compensate by having a very large number of members. I do not know of any biologist who asserts the principle of compensation universally and rigidly.
3. Individuals belong to kinds
   a. From ancient times it was held that kinds of living things are structured in hierarchies of superordinate and subordinate, and as far as I know, the name for a representation of such a structure, namely, a "tree of Porphyry," is the oldest instance of the word "tree" used in this metaphorical way. Porphyry describes the structure but does not diagram it or give it a name. Late medieval logics illustrate it, but the actual name seems to be postmedieval. See Walter Ong (1958, p. 78, fig. 3; from Tartaret 1514), where the tree is an ordinary tree, growing on the earth, embellished with labels and emblems. At its base are individuals; ascending node by node, we finally come to the top of the tree, emblematized by a crown, Substance. (That the more abstract and general is higher, and the more concrete and particular lower, is a metaphysical doctrine of both Plato and Aristotle, long before the tree metaphor was

proposed. Discarded in modern times, it is not relevant to our present story.)

b. Because there is a hierarchy of kinds, each individual thing will (in general) belong to more than one kind. This makes for a difficulty caused by natural language (at least by English and by languages that are like English in this respect): We cannot correctly say that A and B belong to *the* same kind, because there isn't just one such kind, but we cannot grammatically say that they belong to *a* same kind, either. We have to circumlocute awkwardly for example, by saying that there is a kind to which both A and B belong. We would like to say, but cannot, that two organisms of a same kind have a same life cycle, but what we mean is clear. However, the life cycle of a human has more things in common with the life cycle of a dog than with that of a lizard or a lobster, and still more than with a fern or an apple tree. The import of all this metaphor theory is that, when life cycle is relevant for the tenor of "life" taken metaphorically, theorists want to be able to state what life cycle is intended. If the intention is vague or confused or indeterminate, we want to be able to find the vagueness or whatever reflected in unavoidable vagueness of formulation.

c. What concerns us at present about kinds is *change* of kinds. In the eighteenth century, suggestions that kinds of living things change began to appear. A linguist with such suggestions in mind might think of comparing the change from Latin to French not with the change of a living individual from childhood to adulthood, or from prime of life to old age, or the like, but with the change from one taxonomic kind to another, that is, with organic evolution.

d. As regards kinds, one generally assumes (perhaps fictively) that there are lowest kinds, that is, kinds under which there are no subordinate kinds. (Both biology and linguistics will recognize what are called *eschata eide* in Greek, *infimae species* in Latin, understanding the word "species" in the logician's sense, that is, species that have no subspecies. And it is a fact both of living things and of languages that the question of where to stop is a troublesome one to which there is no theoretical answer.) Besides these hierarchically arranged kinds (species, genera, families, etc.), there are other kinds of kinds: (1) Sex: man and many other species of animals are divided into male-kind and female-kind. (2) Social kinds: families ("nuclear" and "extended"), sibs, phratries, moieties, etc. (3) Political kinds: *poleis,* nations, federations, etc. The first kind, sex, is by consensus biological, but over the ages there has been difference of opinion as to whether to subsume the study of social kinds and political kinds under biology.

4. Miscellaneous other properties of living things
   a. Living things need nourishment (food) and other conditions—water, temperature, light. For many lowest kinds, there is a limited range of tolerable conditions; for many others, the range is much wider.
   b. Some living things are useful to us, others are harmful, others are indifferent.
   c. Some living things are more complex or more elaborate than others.
5. Negative properties
   Although linguists have often spoken of languages as young, flourishing, past their prime, moribund, and so on, there are many concepts relating to events and stages of life that they have ignored (e.g., conception, gestation, puberty, menopause), undoubtedly because they are more precise than any biological metaphor that linguists feel moved to borrow. How vague the linguist's biological metaphors are is indicated by the reflection that it is only because languages are female and because they stand in mother-daughter relationships that we can tell whether languages are plants or animals. (In Georg Stiernhielm [1617] languages achieve virility [and so are male]; in Claude Saumaise [1643] they are definitely female; in Andreas Jäger [1686], drawing on both of these, they are both [on Jäger, see below, p. 49; on these, see Metcalf 1974, pp. 236, 249].) It follows that if they are animals at all they are definitely invertebrates. The same consequences follow for all those who by speaking only of mother languages and never of father languages commit languages to parthenogenesis. The point is that negative instances are as instructive as positive ones. We should attend to biological concepts that linguists do *not* use as metaphors, as well as to ones that they *do* use. The former is a larger and much less manageable set.

   Let us now consider the metaphorical uses that linguists have made of these properties.
   That individuals change is commonplace; that species and other kinds change is a theoretical proposition (the doctrine of organic evolution) that has been seriously considered only in the last two hundred and fifty years. When change in language is compared to biological change, what precisely is the term of comparison? Individuals? Species? Until the eighteenth century, when organic evolution began to be seriously considered, only changes in individuals would be considered, either changes within one individual (most notably, the life cycle) or changes from one generation to the next, that is, from parents to children. But in some ways language change is more similar to change in species. It is a complication that biologists themselves, in working out evolutionary theory, assimilated specific change to interindividual changes by

speaking of descent (*Abstammung*) and pedigree. Thus, if even after the rise of organic evolutionary doctrine linguists spoke of mother and daughter languages, they may have only been following the lead of biologists themselves, who spoke (in effect) of mother and daughter species. But there is the notable difference that, whereas change of biological species and change of languages are essentially gradual, generation or production of individuals is essentially discrete, and it is the ignoring of this difference that gives rise to the absurdities pointed up by W. S. Allen (1953, p. 88; cf. Robins 1973, p. 3).

To a large extent, then, linguists have described language change by a metaphor that takes languages as not only individuals and female in sex, but also as parthenogenetic. But when they have turned from a one-parent model to a two-parent model, biological models have proved no more felicitous. With two parents (i.e., with mating), life is subject to the restriction of intraspecific fertility, whereas language is subject to no comparable restriction. In pidgin theory, where the comparison is tempting, linguists have not found any analogue of the biological restriction—far from it.

It is noted above that biologists themselves compared kinds with individuals, not merely in the gross respect that (according to evolutionary doctrine) both change, but also with regard to the fact that a species, genus, or other kind can be said to "rise" or "come into being," wax or flourish, that is, become more numerous in population or diversify in varieties; reach the acme (zenith, peak); "decline;" and "perish" (cease to exist). A good example that has become known in the last one hundred years is the kinds of dinosaurs. Once again, linguists who compare languages with biological individuals are only doing more or less the same thing as biologists themselves have done when they compared biological kinds with individuals. So far as this comparison is regarded as furnishing inductive evidence for a law of nature it is invalid, but so far as it functions heuristically, stimulating inquiry, it can be fruitful. In at least one instance it *was* fruitful: August Schleicher (1821–1868) (cf. pp. 42, 47, 52) should be singled out for special attention as the one who far more than anyone else thought out the comparison between biological and linguistic phenomena and has become notorious for his "Darwinism" (Oertel, Pedersen, Jespersen, Maher, Aarsleff). To interpret Schleicher's pronouncements correctly, a peculiarity of his personality and style must be understood. Hans Aarsleff's interpretation (1982, pp. 294–95; 320 n3), though put forward with great confidence, is poorly researched and largely mistaken. Solomon Lefmann's perceptive "sketch" (1870) remains invaluable, as do the two biographies (1869, 1870) by Schleicher's disciple, Johannes Schmidt. B. Trnka (1952) has shrewd remarks; Rudolf Fischer (1962) newly highlights important facts and supplies a bibliography (prepared by Joachim

Dietze) of writings by and about Schleicher. Patrick Tort's essay (1980) is by far the best recent item on Schleicher's work as a whole.

Schleicher expressed himself with exceptional brevity, force, and clarity, reminding us of Heraclitus and the earlier Wittgenstein, but when he replied to criticism he tended to repeat what he had initially said rather than adding supporting detail or addressing the specific objection that was being put to him. Schleicher stated (and argued for) his views on the nature of languages and the status of linguistic science in 1848, and except for a major change in 1850 (to which he called attention then and later) and another in 1863, replacing (Hegelian) dualism by monism, he adhered to these views until his death. But owing to the peculiarity just mentioned, his several subsequent statements on these matters (at length in 1860, 1863, and 1865 and briefly in 1861) do much less than we would wish to answer the questions that sprang to the minds of his readers. Whitney's criticism that there are blatant contradictions in Schleicher's views was accepted by Johannes Schmidt and by Berthold Delbrück and has become accepted generally (Fischer 1962; see pp. 40, 42 on Whitney).

On the immediate question at hand—are languages to be likened to individuals or to kinds?—Schleicher says in 1848 and again in 1863 that they are like kinds. (Moreover, he holds (1850, pp. 22–23) that a species is a totality of individuals, thus confounding—as almost everyone did before the twentieth century—the relation of class membership with that of class inclusion.) The 1863 essay has this noteworthy statement:

Now Darwin and his predecessors went a step further than the other zoologists and botanists: It is not only individuals that have a life, but also species and genera; these too have come into being gradually. . . . What Darwin established for species of animals and plants, holds also, at least in its principal features, for language-organisms. [My translation]

Schleicher did his linguistic work without self-conscious influence from biology (his hobby, horticulture, has given some scholars the opposite impression); only ex post facto, after Darwinism had made its sensational appearance and at the instance of his Jena colleague Haeckel, did Schleicher (1863, 1865) comment on the parallel between biological and linguistic science. We seem to find, per contra, in the case of Adolphe Pictet (1859) an immediate and self-conscious influence. Pictet (1799–1875), already famous for helping to establish (1837) that the Celtic languages belong to the Indo-European family, published in 1859 and 1863 a two-volume book with the subtitle *Essai de paléontologie linguistique*. He gives us to understand (1.v) that the phrase "linguistic paleontology" is his own coinage. Now in 1844–1846 (second edition 1853–1857) François Jules Pictet (1805–1872), a distant relative of Adolphe Pictet's and like him a professor at the University of Geneva, had published a treatise on paleontology—*real* paleontology—of which Darwin spoke highly in *The Origin of Species.* (Incidentally, in a letter of 3 April 1860 to Asa Gray, Darwin called FJP's review of *OS* "the only quite fair review to have appeared.") Now "Post hoc ergo propter hoc" is not always true, but it is not always false either, and when there is in the mid fifties a treatise on "real" paleontology and in the late fifties one on linguistic paleontology, the two treatises being by two professors at the same university, the hypothesis that the former influenced the latter is not farfetched. I do not suppose that F. J. Pictet's work caused A. P. to undertake his book or that it influenced its contents, but only that it caused A. P. to employ a metaphor, and thus call attention to the analogy between a specialty that had become standard in geology and biology and one that A. P. himself established as a specialty in linguistics.

Adolphe Pictet's book had a momentous effect on Ferdinand de Saussure, which we learn about not from his review (1877) of it nor from his mentions in his *Course of General Linguistics,* but from an autobiography written in 1903, but published only in 1960. When he was twelve or thirteen, Saussure read some chapters of Pictet's book and was (1960, p. 16) "inflamed with . . . enthusiasm" at "the idea that one could, with the aid of one or two syllables of Sanskrit, recover the life of vanished peoples." Somehow young Ferdinand passed from this to the notion that a few radicals, each containing three consonants, were the basis of all actual and even all possible languages. He wrote up his *Essay on languages* (1960, p. 19) in the summer or so of 1872, at age fourteen, and submitted it to Pictet, who was a country neighbor of the Saussure family. Pictet gently demolished the notion, and Saussure's ensuing chagrin over his puerile naiveté explains (as a genetic explanation explains) his unwavering insistence in his mature years that the linguistic sign is arbitrary.

Adolphe Pictet compared his specialty with paleontology, but he owed no more to paleontology than a very general knowledge of it, not any details nor any method. It is quite otherwise when Sir Charles Lyell, at about the same time as Pictet (and as Schleicher's post-Darwinian pronouncements), makes the comparison. This is in his book *The Antiquity of Man* (1863), chapter 23. The position taken in that chapter is all the more interesting for us because here we find a biologist (regarding paleontology as belonging to both biology and to geology) looking to linguistic science for help, not a linguist looking to biology.

There have been two ways of comparing changes involving kinds with changes involving individuals. One of these ways takes these changes as continuous, the other takes them as discrete. If we wish to treat change from kind to kind as continuous, we compare it to growth, that is, to change from one stage to another stage in the life of *one* individual. If we wish to treat it as discrete, we have the alternative option of comparing it instead to a succession of two individuals, one after the other, of comparing it, in other words, to reproduction. When we think of Italian as present-day Latin, we have opted for the continuous model; when we think of it as the daughter of Latin, we have opted for the discrete model. Both models have been popular with linguists.

Now if we conjoin these two metaphors, the result is a "mixed metaphor." And a mixed metaphor is commonly thought to be absurd. But we must distinguish two cases. One is absurd (more specifically, self-contradictory); the other is not, even if it is laughable. A conjunction which is self-contradictory when the conjuncts are taken literally need not be so if they are taken metaphorically, owing to the fact that the tenor of a metaphor is always less than its vehicle. A changing language cannot be both a (discrete) succession of living individuals and a (continuous) succession of states of one living individual, and indeed, we wish to deny that it is either. But it may be *like* a discrete succession and *like* a continuous succession—like the former in some respects and like the latter in others. (And being like a discrete succession in some respects is less than being like it in all respects, i.e. than being *it*.) Likewise with some other mixed metaphors. A language at a certain time cannot be both a boy and a girl, or both an old man and an old woman; but it can be both like a boy and like a girl; and it can even be both more like a boy than like a girl (viz., in respects A_1, A_2, etc.) and more like a girl than like a boy (viz., in respects B_1, B_2, etc.)—provided only that the A-respects do not entail the B-respects nor vice versa. Taking terms metaphorically, or indeed in any nonliteral sense, can change a conjunction that is absurd into one that is merely laughable. (Thus, to say—cf. Kronasser 1952, p. 103, n.87—that blackberries are red when they are green is laughable but not absurd *sensu stricto*.)

A noteworthy subcase is the case where the respect of likeness is farfetched, or fanciful, or trivial. Andreas Jäger (see p. 45) characterized a language as successively an infant, a young woman, a woman in her prime, and an old woman. Why feminine? Apparently for only the most trivial reason, namely, that the word for "language" was feminine in gender. If that is so, then that which distinguishes females from males is no part of the tenor; it would have been more accurate, if pedantic, for Jäger to speak of an infant, a young woman or young man, a woman or man in her or his prime, and an old woman or old man.

Somewhat similarly, when Jäger ascribes to all languages a stage of virility, he may be thinking of *vir* not as opposed to *femina* (or *mulier*) but as opposed to *puer*. If so, his tenor did not include sex, but only relative age.

We can draw two more lessons from Jäger's metaphor. On the same page as where he mixes his biological metaphors, he breaks the life of each language into stages: every language is successively an infant, a young woman, a woman in her prime, and an old woman; and it passes through the stages of infancy, adolescence, virility, prime, and decline and death. Thus, to each language *qua* masculine he ascribes one more stage, not including death, than to that language qua feminine. But since we have found reason to conclude that neither being masculine nor being feminine belongs to the tenor of these metaphors, but that relative age does, we may go on to conclude further that Jäger has given us, in close succession, a four-stage account and a five-stage account of stages in each human life (not counting birth and death) and in each language. But we need not infer carelessness or inconsistency; just as with masculine versus feminine, so here with four stages versus five, the author simply doesn't care about the precise number. His alternation shows not carelessness but indifference.

The last fact of interest to us about Jäger, drawn still from the same page (18, cited by Metcalf 1974, p. 236), is that he treats dialects as branches (*rami*) from a common root. That is, he employs the tree metaphor. His tree is limited to dialects of one language (which we may think of as corresponding to subspecies of one biological species); but it *is* a tree.

Two kinds of jointed structure have been used. The linear, one-dimensional ones are commonly called "chains," the two-dimensional ones are called "trees." Both have joints, or nodes. When one node is immediately connected to two nodes, the resulting structure is redundantly but commonly called a *bi*furcation. A chain contains no bifurcation, a tree contains at least one. Jointed structures of more than two dimensions are thinkable, and in fact a three-dimensional tree would serve to represent language relationship according to the so-called wave theory of comparative linguistics. (A two-dimensional tree would not serve, because if a line connecting two nodes crossed over another such line without intersecting it, that would require a third dimension.) Similarly, it follows from the definition of an "immediate constituent" or "phrase structure" grammar that crossing of branches within its trees is not permitted. Since flexibility is not part of the tenor of the chain metaphor, there are other metaphors (a horsetail, a bamboo pole) that would do as well.

Arthur O. Lovejoy, in his classic *Great Chain of Being* (1936), traces from Greek times to Schelling this immensely influential metaphor. Out of his immense wealth of facts, the two that are of interest here are (1) that the chief scientific uses of the chain metaphor were in biology (biology in the broadest sense, including man) and (2) that the chain was thought of as vertical, with higher (superior) identified with better and lower (inferior) identified with the less good. We should add three points not taken up by Lovejoy. First, in the

Enlightenment, progressivism—a temporalized, literally evolutionary version of the formerly static chain model (Lovejoy, 1936, chap. 9) gave rise (especially through Condorcet, *Esquisse d'un tableau historique des progrès de l'esprit humain*, 1795) to a model that came to be known in anthropology as "the comparative method" (L. H. Morgan, G. J. Frazer, attacked by F. Boas). Second, in the eighteenth century, biologists began considering alternatives to the one-dimensional chain model, notably, trees and networks; see Appendix B. Third, the chain was thought of as starting with Zeus, or with God, and going downward; if one wished to think of moving in the opposite direction, one switched metaphors, and spoke instead of climbing a ladder (*scala naturae, échelle des êtres*). That a ladder and a link chain could serve as different vehicles of the same tenor, unless the tenor itself is to include motion from upper to lower or from lower to upper, shows that various other vehicles would serve equally well, e.g. a series of squares, a series of circles, etc.

The tree model has no such fixed orientation (cf. p. 41). In Schleicher (1853), the opposition of earlier and later in time is represented by the opposition of left and right in space, perhaps by metaphor based on the Western convention of writing and reading from left to right or the convention in Cartesian coordinates (first quadrant), followed in scientific graphs, of identifying the time axis with the x axis when time is one of the variables. (The latter convention may well have been inspired by the former. In the syntactic trees introduced by Chomsky, the root of the tree is at the top. This reflects another Western convention: whereas within one line one moves from left to right, the movement from line to line is from upper [or above] to lower [or below].) But in the article where he first introduced the tree model (Schleicher 1853), the tree looks like a real tree and has its trunk below, its branches above. On the other hand, Darwin's diagram (1936, chap. 4, sec. 8, p. 87) uses an orientation in which below is earlier and above is later, in order—as he himself tells us—that "the horizontal lines may represent successive geological formations, and all the forms beneath the uppermost line may be considered as extinct" (1936, chap. 11, sec. 7, p. 267). This rationale comes from paleontology: In a single series of "undisturbed" sedimentary strata, the earliest (oldest) strata, and hence the earliest fossils, are found at the bottom, and the latest ones at the top.

Notice that although Darwin agrees by and large with the prevailing modern progressivist view that the latter is better, or higher, the rationale of his vertical orientation is the polarity "earlier versus later," not the polarity "less good versus better."

Progressivism flourished earlier in linguistics than in biology. I have in mind those who took the side of the moderns in the Italian *questione della lingua* and over a century later the French *querelle des anciens et des moder-*

nes.[1] Linguists were probably thinking of the evolution of language-kinds well before biologists were thinking of the evolution of plant- and animal-kinds.

Progressivism (and regressivism) need not use the chain model, need not (to use a convenient adjective) be catenary. For instance, one might hold that every one of the Romance languages was better (or less good) than Latin, and yet not hold that of every two Romance languages one was better than the other. And one might use a model that was partly catenary, partly arboraceous (or dendritic). For instance, the comparative method in anthropology (see above) was catenary in relating one main stage to another, and dendritic within each stage.

Such, I gather (from Lovejoy [1936], from Nordenskiöld [1936], and from Daudin, cf. below, pp. 71 – 72) was Bonnet's view of his *échelle des êtres,* and such was Schleicher's view of languages. Schleicher (1848 and all relevant subsequent writings) took over from August Wilhelm Schlegel (1819) and from Wilhelm von Humboldt (about the same time) a classification of language into isolating, agglutinating, and inflecting (see p. 56). Accepting their trichotomy and their view that it was an axiological chain (the inflecting type being the highest), he treated the chain progressivistically, that is, temporalized it: inflecting languages evolved from agglutinating languages, and these in turn evolved from isolating languages. But within the inflecting type (or, as anthropologists would have said, stage) he used—and indeed introduced—the tree model.

Consider family trees. The family tree of a human individual represents some finite portion of his or her blood relationships; in the particular case where only ascending generations are shown, we call the representation a lineage, or pedigree. Human beings have also drawn up family trees of various brute animals; in Western culture, horses and dogs are the most commonly pedigreed; in the Near and Middle East, camels.

When a linguist applies kinship terms and family trees metaphorically to relationships, one would call his metaphors biological. But an important ambiguity lurks here. They are not biological in the sense that the linguist is beholden to biological *science* for them, that is, to biology, for they signify matters known to every normal person, even the least educated. It is not that biological science had to advance to a certain point before they became available to the linguist as vehicles for metaphors.

A glance at the history of kinship metaphors will make this clearer. In the Greek world, the very first use of such a metaphor was by Hesiod (traditionally regarded as the oldest Greek writer other than Homer), in his *Theogony:* Various abstractions, called 'gods,' are placed in a pedigree, which answers the questions which came first, which second, etc. The word 'genealogy,' or rather, the denominative 'to do genealogy,' is first attested in Herodotus, then in Plato and some of his contemporaries (Isocrates, Xenophon). Plato in the *Cratylus* (396b) expressly recalls 'Hesiod's genealogy,' and in the *Timaeus* (22b, 23b) he speaks of (mythical) human genealogies. But by far his most important treatment is in the *Sophist* (226a, 268d), though the passages are not picked up by the dictionary because the specific word 'genealogy' is not used there. Cornford (1935, p. 331) shows how the genealogical metaphor is implicit.

Plato's use of genealogy in the *Sophist* is notable in two respects. (1) The genealogy set forth is not that of an individual human being, but of a type, or kind; not the physical lineage of, say, Protagoras or Gorgias, but the intellectual quasi-lineage of the Sophist in general. So Plato here treats a kind as if it were an individual, just as—two millennia later—biologists come to treat biological kinds as being ancestors and descendents. (2) The two places where Plato genealogizes the Sophist are near the beginning and in the last paragraph of the dialogue. But in the bulk of the dialogue Plato develops at length an entirely different metaphor: The Sophist is the quarry which we are hunting. It is evident that the vehicles of (i) tracing an ancestry and (ii) tracking a quarry are intended to have the same tenor; Plato shifts from the one to the other and back again without a word of explanation.

Although the biological metaphor of kinship is not put to a linguistic use, the fact that the use Plato does put it to is rather like the use made by linguists, and the fact that he here produces so clear an instance of using different vehicles of the same tenor, make it worth mentioning in the present essay.

Here we have in Plato 'genealogy' applied metaphorically but not to anything linguistic. Likewise we do not find in Plato or anywhere in antiquity the word 'organism' applied metaphorically to languages. The metaphor 'organism' differs interestingly from 'genealogy.' We do find this metaphor in antiquity, namely, in the comparison ascribed by Livy (II.32–3) to Menenius Agrippa of the several social classes with the parts of the human body. It was not until the late eighteenth century that biology developed its concept of organism to the point where it began to seem worthwhile to linguists to seek analogues in language. In describing language, antiquity used metaphors drawn not from living things but from artefacts: noun and verb, the fundamental parts of speech, were compared to the timbers of a ship, and the other parts—conjunction, preposition, etc.—to pegs, thongs, and glue.

In all of the examples I have given so far, linguistics has been inspired by biology in one or the other of these two ways. But in addition, there is a third way. Besides looking for analogues to phenomena (like descent) falling under biology, and to phenomena (like organism) which are understood by biology far better than by the outlook of the untrained person, it might try to borrow or to adapt *methods used* by biology. An instance of this third way is taken up in Part II where I discuss comparative anatomy and comparative linguistics.

Of course these three ways are not sharply distinct. There is no natural break in the transition from what everybody knows to specialized science. And as for the difference between the second way and the third, which depends on distinguishing between the phenomena studied and the methods of studying them, it could be contended, even if captiously, that every new method discloses new phenomena and that we can treat the differences between old and new phenomena as primary, and the differences between old and new methods as secondary. But nothing useful would be gained by adopting this contentious viewpoint, because no matter by what viewpoint we describe the situation, inspirations based on what I have called the third way do not culminate in what we would spontaneously call metaphors.

There is yet another reason for enlarging my topic. We must take account of the important fact that it is not only nonbiologists who have used biological metaphors. Biologists themselves have done so likewise, namely, when they have applied the concepts "living," "dead," and the like, not only to biological individuals, but to kinds (taxa), for example when they speak of the taxon Dinosaur and all its sub-taxa as extinct; also when they have given genealogies (pedigrees) of taxa (e.g. Buffon, see Appendix C). A further enlargement of the topic is recognized on p. 63.

But let us review the situation so far. If it was biologists who introduced linguists to horsetails and to bamboo poles (cf. p. 50), it seems that it was linguists who introduced biologists to trees—linguists or at least social scientists. Different pressures are probably the explanation. Biologists were under pressure from the Christian church not to challenge the fixity of species, and they had no unanswerably strong evidence to back up a challenge. Social scientists had much less such pressure, and their evidence *was* unanswerable. Greek writers (Herodotus 1.58, Thucydides 7.57, Strabo 8.333) recognized Greek dialects and the migrations that brought them about. All educated Westerners knew of the differentiation of Latin into the various Romance languages. Schoolmasters and academics were keenly aware that language would change; their Latin or their modern standard French or Spanish or English would change spontaneously if it were not effortfully held fixed. The pressures, then, were opposite: biologists were under pressure to hold that life-kinds do not change, and linguists were under pressure to stop language-kinds from changing.

It was natural, then, for linguists to make use of both models: chain (or bamboo pole) and tree. And biologists could have done this too, even though under pressure not to temporalize. Aristotle launched the chain-of-being idea, and one of his best examples was the biological series: living thing, animal, man. (Daudin 1926, pp. 81–83. Daudin's whole account of Aristotle, pp. 6–

19 and 81–92, is excellent.) But his account had no temporality and therefore was not progressivist. Like the Neoplatonism that elaborated his metaphors and established the Great Chain of Being as standard metaphysics, it had timeless ("static") quasi-processes of states, and real processes of individuals, which being within one state were cyclical. Moreover, Aristotle himself did not use "chain" or "grade" or any metaphor to describe the structure such that some things (plants, the vegetable kingdom) have merely the powers of life, others (brute animals) have in addition the powers of sensation and locomotion, and one (man) has in addition to both these the power of mind. Yet Aristotle himself was the world's greatest student of the varieties of brute animal. What metaphysical status do these varieties have? Do they form a subchain? He left the question not only unanswered but unasked. I call the reader's attention to a now forgotten but valuable work by Mortimer J. Adler (1940) entitled *The Problem of Species*. For example, according to Adler, the problem whether cow is superior to horse does not arise, because cow and horse are not species (in the metaphysical sense) but races (*rationes*), and races are not necessarily subject to catenary order. (Independently of Adler, Leo Spitzer [1948, pp. 147–169] established in convincing detail that the word "race" (as in human race, not as in racetrack) comes from *ratio*.)

Third, Sir Charles Lyell (1863) devoted chapter 23 to a comparison of the origin and development of languages and species. Two linguistic metaphors were debated more intensely and were defended and attacked more vigorously than any other metaphors drawn from biology. These were the two expressed by the words "organic" and "organism." It is commonly but mistakenly thought that the two together form one story. The confusion arose in the very period when debate was active, no later than about 1850, and it was natural enough in a climate where German romanticism and German idealism were so dominant.

Friedrich Schlegel is generally credited with having introduced the comparison of languages with organisms in his *Über die Sprache und Weisheit der Indier* (1808), but anyone who tries to corroborate this claim by actual quotations will be unable to do so. Schlegel does compare the inflected *words* of *one* family of languages (the Indo-European family) with "living germs," and call them organic. In one or two places he calls those languages organic (in a secondary sense) which have *words* that are organic (in the primary sense). I have been unable to find out whether in other writings, or orally, Friedrich Schlegel compares languages to organisms. The new *Kritische Ausgabe* consists of thirty-five volumes so far, not one of which is an index. It is probable that his brother, August Wilhelm Schlegel, contributed to the conception of languages as organisms, but again I have found no firm evidence. Nüsse (1962), though valuable, does not answer this precise question.

We seem to have a hiatus, then, between Friedrich Schlegel's doctrine that *some* languages (those that have inflection) are *organic* and Schleicher's doctrine (cf. pp. 46–47) that *all* languages, in the historic period, are *organisms*. Let us see where we get by working backward. We will find that there is yet a third doctrine to be sorted out: some if not all languages *have* an organism.

Schleicher himself tells us (1850, p. 121; see Fischer 1962; cf. pp. 46–47) that in comparing language-organisms with natural organisms he was preceded by August Friedrich Pott, who says: "It is manifest that, in accordance with the natural history paradigm, one seeks a classification for languages as well. . . . This much is clear, however, that no artificial, Linnaean basis will be of value in linguistics, but only a natural one. . . . Now, in this connection, W. von Humboldt has already accomplished something extraordinary. . . ." (Pott 1840, p. 19, col. A), but (and here Schleicher expressly disagrees with him) Pott adds: "Naturally, language classifications can correspond . . . only very remotely to those of natural history, [because] for instance, any two languages, without exception, are mutually fertile." Roughly speaking, the distance from Pott to Schleicher is the distance from plausible comparison to generally rejected paradox. But before we focus on Schleicher, let us continue working backward. Heyman Steinthal, in his review (1862) of Schleicher (1860), says "That languages are organisms, we have often heard; to what extent they are such, the author hasn't said; but the expression 'function' stems from the organic viewpoint." I wish Steinthal had told us while giving vent to his sense of *déjà vu* what predecessors he had in mind, but no doubt Bopp was one. Bopp wrote in 1827 (quoted by Jespersen 1922, p. 65; I have borrowed Anna Morpurgo Davies's translation, in this volume, with slight changes):

Languages are to be taken as organic natural bodies that form themselves according to definite laws, develop carrying in themselves an internal life principle, and gradually die off, in that, no longer understanding themselves, they shed or mutilate or misuse, that is, they use for purposes for which they were not originally adapted members or forms that were originally significant but that have gradually become a more external mass.

I have not independently "researched" Bopp and draw my data from the valuable chapter by Anna Morpurgo Davies in this volume and from various previous scholars. From all these writings, I get the impression that the organic metaphor here is somewhat casual (which is not the same thing as being fanciful or a *jeu d'esprit*); that, in marked contrast to Schleicher and very like Whitney (Bopp's disciple, cf. p. 42), Bopp regarded the metaphor as appealing and as harmless if not pushed; it could be defensibly used to illustrate an *aperçu*, but not (as Schleicher did) to yield deductive consequences. This metaphor—which is mixed, or at least complex, because it involves a comparison with mind as well as with life—has for its tenor the proposition that no language will last forever but that it will come to an end after some definite time, and gradually, and (more like a higher plant than like a higher animal, be it noted) by suffering progressive damage to or failure of its parts.

To judge from the quotations and citations in the secondary literature, comparison of languages with living organisms is rare in Bopp. In this connection, in the sole occurrence before us, Bopp could not refrain from first modifying the metaphor by another and then returning to the original metaphor, but with an acknowledgment of its awkwardness (*Glieder oder Formen*): These *Glieder,* 'members' or 'limbs', what are they? Branches of a tree? Legs of a cow, or of a centipede? Well, anyway, they are grammatical forms: nominative singular nouns, third singular present indicative verbs, and so on.

Furthermore, Morpurgo Davies says that "there is no confusion . . . [viz., in Bopp] between synchrony and diachrony" (this volume, Chap. 3, note 13), but this is too sweeping. She gives a good example (from 1835), but the example is either unnecessary or insufficient, depending on how one reads the 1827 passage quoted above. For that passage can be read as pointing up the distinction of synchrony from diachrony or as effacing it. It is the phenomenon of gradually dying (*nach und nach absterben*) that allows room for two viewpoints. The language passes from flourishing (acme, let us call it) to utter death, and it passes gradually—which is to say that between these two extreme states there are indefinitely many intermediate ones.

This conception implies a plurality of different states, spread out through time (*dia chronou*) and is diachronic in that literal sense. On the other hand, each of the states is instantaneous, or of negligibly short duration, and the conception of such states is synchronic in that sense. The 1827 passage undoubtedly exhibits the diachronic and synchronic conceptions as I have just defined them, but then *these* conceptions are enjoyed by anyone, even the most ordinary, uneducated person, who has a normal time sense.

However, alongside these conceptions shared by everyone there is a widespread but not universal viewpoint that imposes or finds a value distinction: of all these states, only some one is real (genuine), all the others being apparent (spurious, sham); or all the states form a series such that one extreme is *most* real and all the others are more or less real according to whether they are nearer to or more remote from that one; or in some other way one state is the exemplar, model, or prototype and all the others are corruptions, mutilations, or deficient copies of it. When the serial order is identified with the temporal order, there are two chief versions of this viewpoint: primitivism (the oldest state is the best) and progressivism (the latest state is the best). But there is also a version (I call it acmenism, from the rare adjective *akmēnos* 'full-grown') which holds that things get better, then get worse. This is our prevailing view of Greek culture, for example. It is from an acmenistic viewpoint that languages are most aptly compared with plants and animals, whose life cycles fall into the three main phases of growth, prime, and decline. Then there are viewpoints (collectively called historicism) which say that we have not *explained* or *understood* something until we have constructed a temporal series of which it is the latest term. I suppose one could not be a primitivist, acmenist, or progressivist without being a historicist, but one could be a historicist without being a primitivist, acmenist, or progressivist; Hermann Paul is a case in point (cf. Wells 1947).

The import of Saussure's distinction between synchrony and diachrony was not simply that things change and that change is change from a *state* to a *state*. No normal person needs to be told this. The import was to attack the claim that only the historicist viewpoint is scientific.

Considered in the framework I have proposed, Franz Bopp's view of proto-Indo-European is clearly acmenistic. The 1835 passage quoted by Morpurgo Davies is compatible with that view. Indeed, the fact that "now we do not know that at an earlier stage" of our language such and such was the case is an instance of our present language's shedding or mutilating or misusing something because it no longer understands itself. Whatever Bopp's view of languages viewed all together may have been, his view of the historical, attested Indo-European languages was primitivistic. (That is, they are all of them inferior to the proto-Indo-European language from which they are derived.) As early as 1841 this primitivism was opposed, and progressivism recommended, by the Danish linguist J. N. Madvig. I therefore agree with Hans Aarsleff (1982, p. 300) when he characterizes Madvig as "thus coming close to making the distinction [viz., Saussure's] between diachronic and synchronic study." Or rather, he came closer. He overthrew one error by committing exactly the opposite error. This was an advance, because it brought people closer to seeing how to overthrow the generic error committed by these two opposing specific errors. It was Saussure who first saw the generic overthrow. (Saussure himself may have gone too far. At least people have found it difficult to see how, from his viewpoint, diachronic linguistics has *any* scientific standing.)

The hypothesis that Bopp was either an acmenist or a primitivist is confirmed by his view of analogy (in the linguistic sense of the word). If analogy (more precisely, analogical remodeling) is a monstrosity (*Fehlgeburt*; cf. Bréal 1873, p. xv), then its output must be worse than its input.

As an epilogue to the discussion of Bopp, I would like to sketch the findings of a hoped-for sequel that would carry the episode further. K. F. Becker, a well-known figure in his day, made much of the concept of linguistic organism (Becker 1827). The most famous of all who have in any way applied to linguistic phenomena the concepts of organ, organic, and organism is Schleicher (pp. 46–47, 52) who, beginning in 1850 and most notoriously in 1863 and 1865, called languages organisms outright. Sharp criticism came from two different quarters: the United States (William Dwight Whitney) and France (Michel Bréal and Gaston Paris). Compatriot dissent at first mostly took the form of merely ignoring Schleicher's sensational thesis, although Schleicher's perennial critics, Heyman Steinthal (Berlin) and August Friedrich Pott (Halle), spoke out. But before long rejection was explicit (Friedrich Müller [Vienna]; the Leipzig "neogrammarians," 1876ff.). The French rejection has been recently described, though in unacceptably biased fashion, by Aarsleff (1982, pp. 293–334).

This controversy was nothing but a more general one applied to the special phenomenon of language. The more general controversy is whether—and if so, how—we should distinguish two radically different kinds of sciences: the science of nature and the science of man, or history, or spirit.

In the latter half of the nineteenth century, culminating toward its end and extending well into the twentieth century, a tendency arose to formulate a concept of superorganism, with society as the principal instance. This concept had two sources: philosophy and mathematics. The philosophy of absolute idealism, with Hegel as the most famous contributor, was developed by psychophysicist G. T. Fechner (see James 1909, chap. 4; Merz 1904–1912, 1: 200) in a way that was more popularly usable. Quite independently, the mathematics of probability and of statistics began to develop such concepts as "laws of large numbers" and entropy, on the one hand, and average (statistical mean), on the other hand. These concepts, which were to a large extent developed with the intent of application, found immediate application in physics but also in the social sciences. It is interesting that the pure and especially the applied mathematical work is chiefly French and British. Splendid, distinguished accounts are those by C. C. Gillispie (1963) and John Theodore Merz (1904–1912, vol. 2, chap. 12). Philosophers Josiah Royce (1914, pp. 35–62) and Morris R. Cohen (1936; 1945, pp. 127–154) add valuable insights, but the towering philosopher here is Charles S. Peirce, part of whose life's work was to show that out of chance, by the habit that things have of acquiring habits, there emerges regularity. Peirce and Royce (who was appreciative of Peirce and essentially dependent on him) did all this with perhaps some awareness of the statistical mechanics of Ludwig Boltzmann and J. Willard Gibbs but none of the quantum mechanics started by Max Planck.

With the introduction of statistical and quantum mechanics, we have moved out of biology, and so it is time to leave the topic of organism and turn to another. But it is interesting that we are still dealing with the idea perhaps second most beloved in biology (the first is Purpose), namely, that the whole is more than the sum of its parts.

II

One influence that will preoccupy us is that of comparative anatomy on linguistics. It is of the first importance to keep in mind a certain change that took place in biology just before Georges Cuvier. Nordenskiöld (1936, p. 331) writes about Cuvier's Stuttgart teacher Kielmeyer, who was "influenced by Herder's idea (and Robinet's, see below, p. 70) of a common primal type for all living creatures . . . and who consequently strongly recommended the study of comparative anatomy." Cuvier rejected this doctrine, and even the weaker doctrine that there is a single common type for all animals (*l'animalité*, as Geoffroy St.-Hilaire called it; Meckel also accepted it [Nordenskiöld, 1936, pp. 356–57]). But he did not go to the opposite extreme— Buffon's position, which assigns no two species to the same type. Cuvier's own position—that there are four types: Mollusca, Articulata, Radiata, and Vertebrata—had for its rationale not mere compromise but the claim that nonspeculative observation does not disclose continuity between any one of these types and any one of the other three. Many of the same questions of theory present themselves, whether we are supposing one type for all living things or for all animals or for all vertebrates. Whatever the domain of one type, do we intend to suppose for it a *prototype* or only an *archetype*? If we speak of a source and of derivatives, do we intend our derivation to be in real time or in a merely ideal quasi-time? I can think of only two theoretical questions that are raised by supposing fewer types (say, one for all animals) but not, perhaps, or not as pressingly, by supposing more types (say, four types, as with Cuvier). First, since the rationale for supposing more than one type is discontinuity from one type to another, may we suppose that nonexistent transitional forms bridge the gaps? In particular, may we bring in no-longer-existent or not-yet-existent forms to effect conceptual transitions? Second, as regards any forms that we suppose are possible even though they never have existed and never will, what kind of possibility do we ascribe to them? Real possibility (i.e., that, given the actual laws of the nature of our world these forms might have existed), or possibility in some other sense? And since whether a question is pressing is a subjective matter and a matter of degree, any typology may raise the same questions for one mind as some other raises for another mind.

Comparison in biology is not of interest here for its own sake, but even to deal with it in preparation for dealing with its impact on linguistics we need to grasp some considerations which, as far as I know, have not been clearly set down. The three main considerations are (1) the distinction between archetype and prototype, (2) the distinction between real time and quasi-time, and (3) the initial state that we suppose.

With these considerations in mind, we will see that most discussion of "the comparative method" is highly misleading; that the resemblance between the comparative method in biology (in particular, comparative anatomy) in the first two decades of the nineteenth century and the comparative method in linguistics is superficial and slight; and that mainstream linguistics committed itself unequivocally to prototypes and to derivation (process, development, evolution) in real time a good quarter of a century before biology did (the 1830s versus the 1860s). The best way to begin making these three points is to compare Cuvier with Friedrich Schlegel. This is the usual approach, and even though it is usual, it is the best way.

When we speak of comparison, we always mean comparison in certain respects, whether we name these explicitly or not. There is a fundamental difference between archetypal comparison and other comparison. In the former there is by definition the hypothesis that, given two distinct things (A and B) being compared, there is something C (not necessarily distinct from A and from B, but the case where C = A and the case where C = B are "special cases" and "degenerate cases") from which A is derived and B is derived, something (to use an equivalent terminology) that is the source of A and of B.

It must be noted that "derivation" and "source" themselves have two very different senses: temporal and nontemporal. Again and again, Goethe seems like an evolutionist to post-Darwinians, but he makes it clear that he does not envision evolution in time. The same is true of Karl Ernst von Baer, an especially interesting thinker because whereas Goethe died thirty years too soon for an *Auseinandersetzung* with Darwin, Baer had the opportunity and unequivocally repudiated Darwin's temporalized version. Sir Richard Owen could not do anything unequivocally and cannot be cleared of the charge of bad faith, but after much shuffling he too rejected the temporal version, as did another contemporary of Darwin, Louis Agassiz (Darwin, b. 1809; Owen, b. 1804; Baer, b. 1792; Agassiz, b. 1807). Goethe and his like thought in terms of Neoplatonism, a system worked out from Plotinus (205–270) to Proclus (410–485), which spoke of emanation, return, and so on, but always quasi-temporally, quasi-dynamically, rather than meaning literal time and literal change. The same is true of Hegel, except that there is a new tension. It was possible for Hegel, or at least for Hegelians (as it was not for previous Neoplatonists), to think that *Werden* sometimes involves real time.

Let us speak of "derivatory comparison" ("derivative comparison" would convey an unintended meaning) to mean comparison that seeks, in comparing

A with B, to posit a C from which A and B are derived. The C may or may not be purely hypothetical, or the derivation may or may not be temporal. In the case where C is purely hypothetical and is earlier in time than A and B, it is customary to call the derivation reconstruction. This is particularly true in linguistics, with the result that when linguists speak of the comparative method they invariably mean a certain method that reconstructs a C by means of comparing an A with a B, whereas when anthropologists use the same phrase they have in mind a method whose model is not a tree with at least two branches, but a stalk or stem that is essentially a single straight line. Unless specifically informed of this difference, a linguist will misunderstand if he hears of Franz Boas (1896) attacking the comparative method, as will an anthropologist if he hears a linguist speak of the comparative method as still valid today. It is fair and helpful, then, to say that "comparative" in present-day usage has a sense in anthropology different from the sense it has in linguistics. The two senses are not treated as two species of one genus by either linguists or anthropologists. In order to be unambiguous, I will call the method that is so important in linguistics the comparative-reconstructive method. Comparative reconstruction (it could equally well be called reconstructive comparison) is a species of derivatory comparison. Comparison not of the above-described sort, when done in linguistics, is called "cross-genetic comparison" (C. F. Voegelin) or "contrastive linguistics."

Now we return to biology. Historians of biology (e.g., Nordenskiöld) talk of comparative work—chiefly comparative anatomy and occasionally (as for Haller) comparative physiology—in a sense that includes the linguistic and also the anthropological but that subsumes other comparisons as well. In practice, "comparison" when applied to pre-nineteenth-century work seems to mean the following. Whereas an ordinary anatomy would describe the anatomy of man, the anatomy of the cat, the anatomy of the frog, the anatomy of the carp or the dogfish, one after another, in whatever order, "comparative" anatomy would describe the ear (of man, cat, etc.), the eye (of man, cat, etc.), the skeleton (of man, cat, etc.) in whatever order. The situation is readily understood in terms of a matrix. Let the rows be A, B, C, etc., and let the columns be I, II, III, etc. If we have to represent this two-dimensional matrix by a one-dimensional series, the two most rational methods of doing it are by rows and by columns, that is, AI, AII, etc., then BI, BII, etc., and so on, or else IA, IB, etc., IIA, IIB, etc., and so on. Neither of these two methods of linearizing is more rational than the other. Each method takes one set of sets as primary and the other set as secondary. Seen in the light of this abstract model, the new anatomy called "comparative" is no more comparative than the older anatomy; it merely elects the alternative serialization, taking organs as primary and organisms as secondary, whereas the older anatomy did the opposite.

But before Cuvier a change had taken place. Even the old comparative anatomy was new, compared with ancient and medieval work. But the new comparative anatomy, seeking archetypes, was newer still. The difference between saying (1) "A and B are alike in ways W_1, . . . , W_n and different in ways W_{n+1}, . . . , W_{n+m}" and saying, "Since (1), there must be a C which yields (i.e., from which are derived) A by rules of change R_1, . . . , R_p and B by rules R_{p+1}, R_{p+q}" is somewhat like the difference between synchronic and diachronic linguistics and somewhat like the difference between descriptive and inferential statistics. Inference, reconstruction, and explanation are all related. Cuvier did in biology something like what Rask, Grimm, and Bopp together did in linguistics, and it is a closer analysis of this likeness that here concerns us.

I am aware of changing the topic. The topic "Biological Metaphor" (sc. outside of biology) imposes the restriction that biology must be the source, that we restrict our consideration to ways other sciences or inquiries have been influenced by biology. But in attending to this set topic I have discovered that the restriction is arbitrary, so I wish to broaden it. Two empirical facts that move me are (1) that, as between biology and linguistics, influence has been reciprocal (mutual) rather than unidirectional, and (2) that some likenesses are not lexicalized in metaphors. As writers like to say, the story takes on a life of its own; it dominates me, I don't dominate it. The set topic, then, turns itself into this: likenesses (not necessarily metaphorized) perceived between biology and linguistics, some of them more prominent on the one side, some on the other.

Having introduced the concept of derivatory comparison (and that of reconstructive comparison, one of its species) I turn to models and to relation to time.

The chain model (stalk model), characteristic of what anthropologists call the comparative method, and the branching-tree model, characteristic of what linguists call the comparative method, are not the only models we need to treat. There is a third model, the radiation model. Arsène Darmesteter introduced it in his *Vie des mots* (1886) as a model for relating the various senses of a word. One takes one sense (perhaps hypothetical) as central and treats each of the remaining senses as derived from the central sense. In Darmesteter's semantical use of the model (taken up by James B. Greenough and George L. Kittredge [1901, Chap. 18, esp. p. 261] and by Roman Jakobson in his concept of *Gesamtbedeutung*), the model is derivatory and becomes inferential if the condition is added that the central sense is treated as merely hypothetical, and it becomes reconstructive under the additional condition that the central sense occurs earlier in time than any derived sense. (Note that, as also in reconstruction in the style of linguistics, one may treat the actual as merely hypothetical. For example, by reconstructing a language from which all the present-day Romance languages are derived, one arrives at the hypothesis of a language that is actually very much like the Latin we know directly from records.)

What has gone before deals in effect with that part of linguistics called historical (diachronic), and especially with historical work that makes use of "the comparative method." It accepts without question the prevailing view that Cuvier's comparative anatomy was a major influence on comparative linguistics, furnishing inspiration to it and supplying a model. The time has come to question that view, to treat it not as an established fact or dogma or principle but as a hypothesis that is subject to reexamination at any time. And indeed, as soon as we start questioning this hypothesis we find much to question.

It turns out that the principal evidence for the hypothesis is that everybody asserts it. But a hypothesis is tested by its consequences. That p is widely believed to be true is only one of the consequences of p's being true, and by no means the surest one. Cuvier's "comparative" work was not the only exciting "comparative" work being done in his day. The likenesses between plants and animals (such as palm trees and dinosaurs) and languages are few, and the notable differences are many. To give the briefest possible version of what happened, the only notable likeness between what Cuvier did and what Bopp did is that both, from things in existence today, drew inferences about things no longer in existence today that existed long ago. I emphasize that in drawing *his* inferences Cuvier relied essentially on his "principle of correlation," and that Bopp, in drawing *his* inferences, made no use of any linguistic counterpart of that principle.

It is part of the traditional account that the quickened response was first displayed in the classic work of Friedrich Schlegel, *Über die Sprache und Weisheit der Indier* (Heidelberg, 1808). But this item is as fragile as the traditional account of which it is a part. Friedrich Schlegel (we must distinguish him from his older brother, August Wilhelm) does allude to "comparative anatomy" as a model and does himself make quite a number of comparisons between Sanskrit (= old Indic) and the Indo-European languages, more familiar to Occidentals, namely, Greek, Latin, German, and occasionally others. But all this is only the barest allusion to a likeness between the biological facts and the linguistic. As for inferences, Friedrich Schlegel does not so much as raise the question whether Sanskrit was itself the source of these other languages or whether there was some no-longer-attested language—the one now called proto-Indo-European—that was the source (so that, to use the terminology mentioned above, he thinks of an archetype but not of a prototype). Moreover, although Cuvier and the comparative linguists both drew inferences and advanced hypotheses, Cuvier never (to my knowledge) advanced inferences about whole organisms (e.g., never predicted archaeopteryx and archaeornis), but only about missing parts of organisms that partially survive at present (chiefly their bones, as fossils). Linguists, per contra, rarely fleshed out the products of their inferences, but freely inferred whole languages. Thus the likeness (analogy) between the facts of life and the facts of language, and between the work of comparative biologists and comparative linguists, diminishes on scrutiny.

In dealing with biology in relation to linguistics, we must take into account not only changes in individuals and in species but also changes in views, concepts, conceptions, ideas, theories, and theses, so that without any intention of being clever or cute we find ourselves speaking of the development or the evolution of comparative methods. That this is not mere *jeu d'esprit* or a *façon de parler* is shown by its consequences, in particular by our discovery that the development is gradual.

So far, it has been pointed out that well before Cuvier's time people spoke of comparative anatomy and that, on the other hand, Cuvier intended something by this phrase that his predecessors did not intend. Cuvier is said to have done this, that, and the other, but if we look at what he did, not in the way of details but in the way of general principles and general methods, we find that very little is new. Even his famous Principle of Correlation, far from being new, is often appealed to by Aristotle (as is noted passim by William Ogle in his annotated edition [1882] of *The Parts of Animals*).

Rather than conclude that Cuvier's common reputation is baseless, we should enlarge our approach. Under this enlarged approach, we have no difficulty explaining Cuvier's impact. He brought to biology enormous vitality, unflagging and well-channeled; great articulateness, style, and gifts of popularization; the extraordinary good luck to be making spectacular finds in the very ground on which his Parisian audience lived; and leadership qualities and administrative ability. It is also in his favor that he not merely commenced many great projects but completed them. He was undoubtedly one of the great scientific figures of all time, and my purpose here is not to reevaluate him but only to consider his possible influence on linguistics.

Cuvier was not the first to recognize and formulate the Principle of Correlation, but we must add that he was the first to *use* it in a certain way. There was great interest in fossils in the eighteenth century, as explained in Greene (1961, chap. 4). By 1800, people were beyond merely asking which animal the fossil bones belonged to and were willing to consider the possibility that they might belong to some now extinct animal. It was in this situation that Cuvier proposed to use an old principle in a new way, namely, to draw inferences about what the rest of the animal was like when we have only some bones, in other words, to contribute toward *reconstructing* the animal.

Our understanding of Cuvier is advanced by Huxley's incisive essay "On the Method of Zadig" (1880; reprinted in vol. 4 of Huxley's *Collected Essays* [1893–1895]). Cuvier had claimed to offer "une marque plus sûre que toutes celles de Zadig" ("a mark surer than all those of Zadig", alluding to a story by Voltaire). Huxley shows brilliantly that Cuvier gives a sound method but an unsound account of its rationale. When people go so far as to speak of Cuvier's new comparative anatomy (e.g., Gillispie 1951), we gather that one of the chief things they have in mind is its inferential side, surer than Zadig: its claim, about which Cuvier grew bolder as the years passed, that from one part of a fossil animal one can "reconstruct" other parts—in the boldest version, all the other parts. Huxley shows that this is not true even in paleontology. The fact is that Cuvier was both very bold and very lucky. His exceptional boldness was to a degree justified by his exceptional memory (Sneath 1964, pp. 480–481), but still he was also lucky. Huxley (1893, pp. 18–19; cf. Desmond 1982, pp. 57–8, 80 on Huxley's opposition to Cuvier) gives an example of a spectacular prediction: A piece of rock, split into two parts, revealed a fossil. The two pieces were brought to Cuvier. On examining the teeth and the lower jaw, which were visible at the two exposed surfaces, Cuvier predicted a certain unusual feature of the pelvis. While others watched, the hidden pelvis was brought into view, and it had the unusual feature he had predicted. The example simultaneously illustrates (1) that Cuvier's prediction had no basis in physiology, (2) that it did have a basis in observable correlations of which Cuvier had an enormous store in his memory (from the teeth and jaw he had recognized the animal as an opossum), and (3) that soon many correlations of the form "if A then B" became known which would not afford a probability close to 100 percent but only an either/or. And now I may give my reason for questioning the prevailing opinion that Cuvier influenced the thinking of Friedrich Schlegel's *Sprache und Weisheit der Indier*—at least we may question whether there was any specific, pinpointed influence, notwithstanding the circumstances that Schlegel was studying Sanskrit in Paris during years (1802–1808) when Cuvier was the talk of the town (Aarsleff 1982, pp. 33–34) and that Schlegel negotiated a letter of recommendation from Cuvier (Aarsleff 1967, p. 155 n 108). The reason for questioning Cuvier's influence on Schlegel is just that although Schlegel prominently calls for a comparative grammar "like," (i.e., modeled on) comparative anatomy, what he actually produces is nothing of the sort. If we do with Schlegel what Huxley did with Cuvier—compare what he says with what he does—we find that he does nothing distinctively Cuvierian. Schlegel advances a little step, but not a big one, beyond Sir William Jones (1786) in opening up the possibility of a now extinct parent language, but though he *speaks* of a comparative grammar analogous to comparative anatomy, he offers no close analogue to a Cuvierian reconstruction, which would have been a restoration of all or much of some extinct language from surviving fragments of it. He did not even offer what would have been a looser analogue to Cuvier but a closer adumbration of subsequent linguistics, a reconstruction of a *wholly* extinct language from surviving *descendants* of it. All that he did, as regards the comparative-reconstructive treatment of Indo-European, was establish by a wealth of examples that Sanskrit, Greek, Latin, and Germanic are somehow closely related, that is, that there *is* an Indo-European family. But this only did in greater detail what had been done before him by Halhed. (Lord Monboddo and Sir William Jones are derivative from Halhed. We may disregard Fathers Pons, Coeurdoux, and Bartholomew of St. Paul, not because they did not do anything but because it was not published or divulged until much later). After all, Friedrich Schlegel's book is about the language and wisdom of the Hindus, not that of the Indo-Europeans.

The only influence we can find of Cuvier on Friedrich Schlegel is a very general one: by careful study of what we have, we can draw inferences about what once was but what we no longer directly have. This is reconstruction in a very broad sense, but still it gives rise to a project—or a research program, in the sense of Imre Lakatos—such as the world had not seen before.

But Cuvier actually did something else very relevant to linguistics—and to its resemblance to biology—that is usually presented as a matter not connected to his comparative method. Loren Eiseley (1958, pp. 117–19) quietly but decisively notes the crucial idea: Cuvier's "insistence upon unrelated structural types . . . was a necessary preliminary to the kind of branching evolutionary phylogeny which is now everywhere accepted" (p. 118).

Cuvier rejected the chain of being (*l'échelle des êtres*). Whereas a major strain of eighteenth-century French biological thought endeavored to fit all living things, or at least all animals, into one chain, Cuvier set up four "types," which he treated as unrelated to one another. (Others tried modifying the chain model in various ways: recognize two chains, one for plants, one for animals; or use instead of the one-dimensional chain model a two-dimensional model [Linnaeus' map; Haller's net; cf. p. 72]. I do not know of any case, until after Cuvier's time, when anyone proposed a tree model.) Cuvier's model was essentially and intentionally negative. It was not merely that at the present time one did not see how to relate the four type-plans to each other; relation was denied. But it did a service in stimulating the consideration of other models than the linear one. And it strikes me that Friedrich Schlegel is Cuvierian in this negative way. He has indeed a scale of language-types from worst to best, in the old-fashioned way, and he places Indo-European in the highest type. But the series is merely ideal, not temporal, and he emphatically denies that the highest type—that of inflecting languages—could have evolved from any other type. Thus it was unrelated to the other types in a sense quite analogous to Cuvier's.

Appendix A: *Kant* (cf. p. 39)

In two pages of Section 80 of the *Critique of Judgment* Kant pours out a dazzling succession of insights. First, the enterprise of explaining mechanistically "natural products" that show design (Kant has in mind, as he tells us on p. 268 and cf. Sec. 58, p. 194, (a) living organisms and (b) crystals) is praiseworthy though impossible. The natural scientist must make the attempt. Second, Kant explicitly mentions (p. 267) comparative anatomy and its search for a system. Third, the comparative anatomist notes the presence of a "scheme" (*Schema*) common to many animals, such that "with an admirable simplicity of original outline (*Grundriss*) a great variety of species has been produced by the shortening of one member and the lengthening of another, the involution of this part and the evolution of that"; and "this analogy of forms, which with all their differences seem to have been produced according to a common original-type (*Urbild*), strengthens our suspicions of an actual relationship between them in their production from a common parent, through the gradual approximation of one animal genus to another . . ." A footnote to the following paragraph considers this "daring venture of reason," namely, "A hypothesis . . . [which] considers one organic being as derived from another organic being, although from one which is specifically different; e.g., certain water animals transform themselves gradually into marsh animals and from these, after some generations, into land animals." Fourth, Kant mentions (p. 268) raw matter, mosses and lichens, polyps, and man; that is, a linear *échelle des êtres*—the catenary, not the dendritic model.

Comments. (1) In Kant's system, it is no paradox that we should try even though we cannot succeed. Trying is a matter of Reason, succeeding of Understanding, and the contrast between constitutive principles of Understanding and regulative principles of Reason is fundamental to Kant's Critical Philosophy (see especially *Critique of Pure Reason*, Section v.i), even though most surveys of Kant slight it or even reject it.

The claim that we cannot explain purpose (nor, therefore, life) mechanistically is expressed by Kant in this dramatic way: There will never arise a genius who can explain (mechanistically) even a grassblade as Newton explained the inorganic cosmos (Sect. 75 end, p. 248; Sect. 77 end, p. 258). (Compare Cassirer 1950, pp. 163, 211.)

(2) Note that in Kant's phrase (p. 267) "comparative anatomy," he uses the Latin word *komparativ*, not the German word *vergleichend*. He does not, however, draw any distinction between these two synonyms. Sometimes Kant treats a Greek or Latin word and a German word that is more or less synonymous with it as sheer synonyms (e.g., *Archetypon* and *Urbild*, Sect. 51, p. 166); sometimes he has a nuanced distinction, as between *Evolution* (Sect. 81, p. 272) and *Auswicklung* (Sect. 80, p. 267) both of which are naturally rendered "evolution" in English; in the absence of any distinction, express or implied, in the *Critique of Judgment* between *komparativ* and *vergleichend* one may infer that Kant intends none.

(3) The reference to shortening and lengthening makes us think of the various distortions described in Thompson (1942). But the idea so beautifully worked out by Thompson (and, in one instance, by Darwin; see Ghiselin 1969, p. 127 and n. 60) can be traced back to Peter Camper (1722–89), who, as Huxley notes in his 1854 article, showed how to convert a human arm into a bird's wing, etc., by altering proportions. Indeed, a particular instance— namely, the "facial angle" of the human cranium—is, perhaps, Camper's best-known accomplishment (see Greene 1961, pp. 190–94 on Camper; p. 193 illustrates from his posthumous treatise the grids that anticipate Thompson). We can go back further. Camper mentions, as idealizations of the human head not to be found in nature, certain "heads modelled by the ancient Greek sculptors" (Greene's words, p. 192; cf. his quotation on p. 194 from Camper p. 22). Now grids very like Camper's and Thompson's are to be found (Thompson p. 1053; it is surprising that though Thompson mentions Dürer he does not mention Camper) in Albrecht Dürer; a convenient reproduction is to be found in Kunz (p. 303). Returning to Camper, I note that Kant mentions him twice in the *Critique of Judgment* (Sect. 43, p. 146; Sect. 82, p. 278, though not on this matter); and moreover that one of Kant's examples of objective purposiveness in nature is (Sect. 61, p. 206) that birds' bones are hollow [to make flight easier], which implies knowledge of Camper's work if Nordenskiöld is accurate in his assertion (p. 260) that Camper, in his "study of the bone-structure of birds, . . . describes for the first time how the bones are filled with air".

So Kant may have Camper in mind, although I have not found the detail of the connection. But there is another possibility. Already in 1754 (text in Lovejoy, 1936, pp. 278–79), Diderot had written: "When one . . . observes that, among the quadrupeds, there is not one of which the . . . parts . . . are not entirely similar to those of another quadruped, would not one readily believe that Nature has done no more than lengthen, shorten, transform, multiply, or obliterate, certain organs? . . ."

(4) Diderot goes on to consider the possibility that all beings have one

common prototype, a possibility turned into a thesis by J. B. Robinet (1735–1820) (Lovejoy, 1936, pp. 269–83; cf. above, p. 000, Herder and G. St.-Hilaire). Just as important for Robinet as his single-prototype thesis is the *échelle-des-êtres* thesis: The principle of continuity orders all living things into *one linear series*—just one array, not two or more, and that array a linear (catenary) one. Kant in Section 80 entertains this as a hypothesis. In the *Critique of Pure Reason* (as noted by Lovejoy, though with wrong references; the reference is 668/696, i.e., p. 668 in the First Edition [1781], p. 696 in the Second Edition [1787]), Kant accepts the principle as a regulative idea but not—unlike Leibniz—as constitutive.

(5) Here are places in the *Critique of Pure Reason* where the constitutive-regulative contrast is discussed: 179/221–2, 236/296 ff; 509/537; 515/593 ff; 642/670 ff; 646/674 (concepts never realized, but systematically necessary); 664/692; 693/721 ff.

The same distinction also figures prominently in the *Critique of Judgment*. There is no standard pagination of this work as there is for the *Critique of Pure Reason* (viz., that of the 1781 edition and that of the 1787 edition); the nearest thing to it is that more recent German editions indicate the pagination of the third German edition (Berlin: Lagarde, 1799). I will give first the 1799 pagination and then its equivalent in Kant (1951). Here, then, is the list: pp. iv–v = 3-4 (Preface); 260=200 (Sect. 59, end); 270=207 (Sect. 61, end); 294=222 (Sect. 5, end); 301=226 (Sect. 67); 339=249 (Sect. 76); 342=251 (Sect. 76); 344=253 (Sect. 76, end); 429=304 (Sect. 88); 437=309 (Sect. 88). Ernst Cassirer 124-6, 133-4 is helpful.

(6) Goethe acquired the *Critique of Judgment* within half a year of its publication and we have his copy, with its underlinings and marginal markings. Vorländer 1959 gives a useful précis (XXV-XXX, = Sect. V of his Introduction). Goethe found Section 80 especially interesting (XXIX).

Appendix B: Pallas

Peter Simon Pallas (1741–1811) is highly thought of as a zoologist, and very poorly thought of as a linguist. It is not generally known that he was a zoologist by his own choice, and a linguist by the choice of his patron Catherine the Great—a "choice," that is to say an insistence which, since it was imperious, he was not in a position to refuse.

In considering interactions and parallels between biology and linguistics, it is natural to formulate, as one of our many tasks, the task of looking for individuals who were both biologists and linguists. The number turns out to be small indeed. To confine ourselves to the Renaissance and later, Conrad Gesner, the founder of bibliography, was equally important in linguistics; for he founded also a genre of monograph, the Paternoster-genre, in which one text is presented in two or more languages. And thirdly, he also produced "four immense folio volumes" (Nordenskiöld 1936, p. 93) of natural history. Olaf Rudbeck (1630–1702) achieved distinction in anatomy (Nordenskiöld 1936, pp. 145–47) and in linguistics (Metcalf 1974, Diderichsen 1974). Even after the age of specialization had begun, S. L. Endlicher in his short life (1805–49) made a name for himself both in botanical taxonomy (Nordenskiöld 1936, pp. 438–39) and in Sinology, not to mention Celtic studies. Pallas, in contrast to these, found himself engaged in linguistics involuntarily.

This significant but little-known fact we have from Klaproth's testimony, picked up by Wiseman, pp. 27–28. It is significant because it explains the qualitative difference between Pallas's innovative zoology and his old-fashioned and poorly executed linguistics, almost the last of those Gesner-style Paternoster-collections whose history is recounted by Wiseman, p. 29. The word "comparative" in his title must not mislead us; as Pott (1883, p. 295) observes, the comparison is "indeed a vastly different sort of thing from Bopp's." To be positive and specific, it is what C. F. Vogelin called cross-genetic comparison, though not of the variety called by Robins and by A. Morpurgo Davies typological. Gulya gives some detail on the outcome.

Of interest to us here is not Pallas the reluctant linguist but Pallas the precocious naturalist. His book, *Elenchus zoophytorum* ("An inventory of the zoophytes"), has a twofold aim: theoretical and descriptive. The descriptive aim, to which the bulk of the work (pp. 25–432) is primarily devoted, is to take account of all that is known at the time of writing about the zoophytes and some discussion of questionable cases; the theoretical aim—expounded primarily on pp. 3–29 (and see also the "Preface", pp. vii–xvi) is to show that (a) the zoophytes form a natural order, which is (b) characterizable by their having both animal and vegetable characteristics (hence their name), (c) arrangeable in a linear series, and (d) forming a group intermediate between sheer animals and sheer vegetables. Naturally the theory and the description penetrate each other.

At the time when Pallas was writing, the Great Chain of Being was very popular. Restricted to biology, the doctrine holds that all living things can be ordered in a single linear series, from (though of course not including) nonliving things at the bottom to the highest animals (man excepted) at the top. The leading advocate of this doctrine was Charles Bonnet (1764); Pallas says (p. 3, fn., i.e., at the beginning of the exposition of theory) that what Bonnet teaches in his recent distinguished book does not differ very much from Pallas's own views. Daudin (163, n.2) discounts Pallas's acknowledgment, on the ground that the substance of Pallas's conception and the nature of his work show clearly that Bonnet was in no degree the source. What is at issue here is the question how different Pallas's conception was from Bonnet's.

Pallas expounds his conception in one paragraph, at the end (pp. 23–24) of his exposition of theory. He considers four possible structures in which all nonhuman living things might be placed: a one-dimensional chain or ladder; a three-dimensional polyhedron; a two-dimensional net; and a two-dimensional tree.

(1) Chain/ladder. Daudin's outstandingly good history of biological classifications (in the half-century 1740–1790), cited by Lovejoy (1936, pp. 61, 225) on other matters, has in Chapter 2 a brilliant discussion of Bonnet.

(2) Pallas says the genera of organic bodies could be better disposed on the faces (*areolae*) of a polyhedral figure with many compartments (*figura polyedra, multilocularis*) than in Bonnet's ladder. The suggestion appears to be his own; he pursues it no further, but turns to

(3) The net. This model he ascribes to Vitaliano Donati, whom I shall quote. In a monograph (cast in the form of a letter dated 1745) published in 1750, Donati says, at the beginning of Chapter 4:

In all these orders, in all these classes, nature forms her series and passes insensibly from one link of her chain to another. Moreover, the links of one chain are interlaced with the links of another chain in such wise that one would have to compare the progressions of nature rather to a network (*filet à réseau*) than to a chain. It is a tissue of many threads, which communicate with, relate to, and unite with each other. (Translated from the 1758 French translation, p. 20.)

It is true, as Daudin notes (1926, p. 114, n.2) that Donati no sooner introduced the net metaphor than he dropped it in favor of the old-fashioned chain/ladder; but nevertheless his metaphor left its effect. Besides two very good pages (pp. 111–112) on Donati himself, Daudin devotes a whole section (Sect. 3.3 = pp. 159–73) to the net metaphor. I can add two more instances to those he gives. (1) Albrecht von Haller (1768, 2.130), dealing with Gladiolus (in the Iris family), which is affine both with the Lilies and the Orchids, makes this passing remark: "Nature has connected her genera in a net (*reticulum*), not in a chain; mankind cannot follow anything but a chain, because it cannot in speech expound several things at once." (This passage is quoted in Hanson [1958, p. 69] without indication of secondary source and without indication that it is an obiter dictum on one genus, not a general thesis.) (2) Linnaeus uses a map metaphor in a way that is more or less equivalent to the network metaphor; see Appendix D.

(4) But, says Pallas, the system of organic bodies would be best of all outlined by the image of a tree, whose root immediately puts forth a double trunk: animal and vegetable. He says no more about the vegetable trunk (beyond the remark that in plants we can find natural orders but not, as we can in animals, natural classes), but describes the animal trunk as follows: It goes on

up from "molluscs" (a subdivision of Linnaeus's "worms"), a class of which zoophytes are one order, to fishes, but with a sidebranch between these going off to insects; from fishes to amphibia, and then, at the tip, to quadrupeds; but again between amphibia and quadrupeds there is a sidebranch, going off to birds.

Now we can answer the question how close Pallas's own conception is to Bonnet's. Daudin himself makes it clear (pp. 161–63, cf. above p. 71) that Bonnet is aware of difficulties with the ladder, and (p. 162, n.2) is prepared to admit, sometimes, "lateral and parallel" branches. May we not then suppose that Pallas took the essence of Bonnet's view to be continuity (no leaps in nature, and no empty places), not linearity, so that to admit bifurcations and lateral branchings as freely as Pallas did would not be an essential difference from Bonnet?

Daudin (p. 166) thinks that Pallas preferred the tree to the polyhedron just because it was simpler and equally sufficient. Johann Hermann, seventeen years later, described in detail a three-dimensional model (pp. 170–71).

One last point of interest about Pallas's proposal. Whether because he denies the continuity from ape to man, or merely refrains from affirming it, or (an even more noncommittal attitude) considers man outside the purview of the zoologist, the tree described by Pallas does not include man. (Neither does Bonnet's ladder: Daudin p. 178.) At the other end—here is the point of interest to us—, he definitely denies continuity; we may use Richard's concepts of vehicle and tenor to describe what Pallas does. Calling certain structures "trees" had been done before him (e.g. by Buffon 1755; see Appendix C) and became standard after him; but he alone, as far as I know, included in the tenor of this vehicle the thought that the "tree" is *dis*continuous with its base (p. 24; noted by Daudin, p. 180, n.2). The root is not continuous with the soil as it is continuous with the trunk; it merely is situated on and in soil. The purpose of this unusual departure from the ordinary tenor is polemical. Linnaeus (1744) had divided all substances into three coordinate (though ranked) kingdoms: mineral, vegetable, animal (Daudin pp. 177–88). But though Linnaeus's intention was good—to set up a ladder—, his trichotomy was rejected by Buffon (1749), Pallas (1766), Blumenbach (1779), Daubenton (1782), Lamarck (1778), Vicq-d'Azyr (1786), A. L. de Jussieu (1789), and Desfontaines (1795). All agreed that it should be replaced by a dichotomy – inorganic and organic, or nonliving and living, and then the organic subjected in turn to dichotomy. Neither Pallas nor—to judge from the information in Daudin— any of these others remarked that while ladders are incapable of representing the difference between Linnaeus's conception and theirs, trees do it easily: a root with three trunks, versus a root with two trunks of which one but not the other divides into two branches.

Appendix C: Buffon

Greene (1961, p. 152) brings to our attention a remarkable genealogical tree published in 1755 by Buffon. It is so important for the history of the tree metaphor that I give a minute discussion of it. Greene's widely available reproduction is accurate except for details which I report below. In order to fill a gap in Buffon's terminology by concise and familiar words, I have introduced the words "ancestor, descendant, parent, child" into my description (though not into quotations from him).

As is usual in accounts that make use of genealogical trees, Buffon mixes metaphors literally appropriate to trees with terms appropriate to blood relationships among human beings. Moreover, as is usual in genealogical trees where the items are not biological individuals but biological taxa (in the present instance, races of dogs), words literally signifying blood relatives or blood relationships are themselves used metaphorically. (It is a sign of metaphor that although he speaks of "mongrel" races, no meaning is attached to identifying one of the parents as male and the other as female.) This is my warrant for adding the relative terms "ancestor, parent, descendant, child", and the relationship term "to descend from, be descended from, be a descendant of", although he himself does not use any of these. (Instead of saying that A is the parent of B, Buffon says that A degenerates into B; the import of this term is well discussed by Daudin, 1926, pp. 139–42.)

Buffon says: "In order to give a more exact idea of the order of dogs, of their degeneration in different climates, and of the mixture of their races, I here subjoin a table, or, if you like, a sort of genealogical tree [*une espèce d'arbre généalogique*], in which all three varieties can be seen at a glance; this table is oriented like maps, and so far as possible the respective positions of the climates have been preserved" [my translation; Buffon 1812, p. 391 is a little less literal].

Buffon's "table" is more complex than any we have considered so far, and that is why I have deferred the account of it until this point, although it is earlier in time (1755) than the others. It assumes the burden of reporting all that ordinary genealogical trees report and more; the added burden accounts for two differences from the familiar style of genealogical tree. First, an ordinary tree will self-consistently show either every item represented on it as having one parent, or else every item as having two. But Buffon is dealing with the intraspecific case, where there are races, call any one of them A, such that A has only one parent breed, and then other races, B, such that B has two parents. In the first subcase, B results from its parent by degeneration, artificial selection, or the like; in the second subcase, it results from its two parents by mongrelization (hybridization). But the trees we have hitherto considered only provided for single-parent items because only taxa higher in rank than species were items for them, and it was generally agreed that only taxa lower in rank than species—only races (subspecies, strains, breeds, varieties), in other words—can be mutually fertile. Second, Buffon wishes to have the spatial disposition of the races in the table iconize, as far as feasible, their geographical distribution. (Hence the arrow showing that up represents north, following the convention of Occidental map-makers.)

As actually laid out, the table conforms rather poorly to the second stipulation. The north-south difference is pretty well reflected, but not the east-west difference. British and Danish breeds are found represented not only far to the viewer's left, as they should be, but also far to the right. It would have been possible to draw the latter close to the former, and make the other changes entailed by this, but it would have spoiled the pleasing balance, for it would have much clustering on the left and much empty space on the right. We learn from this consideration that the explicit stipulation can be overridden by an implicit demand for pleasingness and perspicuity.

The table consists of shapes and lines. The shapes are four hexagons and thirty-three circles. The largest hexagon—about one third larger in area than the other three—represents the first ancestor, viz. the shepherd's dog from which all other races of dog descend, immediately or mediately. A shape representing race A is connected by a line to a shape representing race B in order to represent that A is a parent (immediate ancestor) of B, or vice versa: a solid line if it is sole parent, a dotted line if it is co-parent.

If a race has at most one ancestor and at least one descendant, it is represented by a hexagon; if it has two or more ancestors, or no descendants, or both, by a circle. If a circle representing race A and a circle representing B are connected by a solid line, one cannot tell which of the two is the parent of the other; one must follow the lineage further. If the line is dotted, then either A or B must touch a second dotted line, at the same point where it touches the first line segment. Thanks to this convention we know that if a dotted line forms at A's circle but not at B's circle a vertex with another dotted line, then A is the child and B is a co-parent of A.

With one or two exceptions, Buffon does not formulate his rules; it is up to us to figure them out by comparing his shapes and lines with his descriptions in words of the races represented by the shapes and their relations represented by the lines. At several points rules other than the ones I have formulated, and equally plausible, would fit the data as well.

In addition to solid and dotted lines, Buffon employs a third kind of connection: adjacency. The four cases where two circles are merely tangent to each other are: (1) Pyramus and King Charles's dog; (2) large spaniel and small spaniel; (3) terrier with crooked legs and terrier; and (4) harrier of

Bengal and harrier. Buffon's explanation of (3) and (4) make me think that
adjacency is meant to represent *slight* differences. (He says in the same paragraph [1812, p. 392] that the terrier, the hound, and the harrier are proved to
belong to the same race by the fact that a certain litter contained all three of
them.) It would seem that in the last analysis, his table represents not what he
regards as races, but what common opinion regards as races. It seems inconsistent to say on p. 393 that the great Danish dog derives from the Irish
greyhound, and on p. 394 just the opposite, but on Buffon's climatic theory the
Irish greyhound transported to Denmark would become the great Danish dog,
which if it were taken back to Ireland would become the Irish greyhound
again, so that it is arbitrary which one we call parent and which one child.

Such is the table of 1755. The reproduction in Buffon (1812), which reduces the size, translates the French names into English, and omits the references in Buffon (1755) to the plates depicting the several races, is otherwise
accurate except for making all four hexagons the same size. Greene (1961,
p. 152), a redrawing of the 1812 plate (thus inheriting the inaccuracy of the
hexagons), makes the plate higher than wide (in Buffon [1755] and Buffon
[1812] it is wider than high) and omits the northward-pointing arrow in the
lower right corner. (The original, 1959 edition does not make this change.)
The high-wide distortion has no importance, but leaving out the arrow was a
significant change. Greene's caption calls the table "the derivation of various
races and breeds of dogs from the wolf dog by artificial selection". Buffon—
at least in pages explaining this table—draws no distinction between races
and breeds, says nothing of artificial selection (his processes are climatic degeneration, hybridization, training, and inheritance of acquired characteristics), and expressly and repeatedly names the shepherd's dog, not the wolf
dog, as the trunk (*souche*) of the genealogical tree, i.e. as (in my terminology) the first ancestor. My reason for mentioning these details is that otherwise the reader would be puzzled by the discrepancies between my account
and what he finds in Greene p. 152.

Buffon's table does not look like a tree because of the geographical information with which he encumbered it. I call it an encumbrance because he
could not do it thoroughly without loss of perspicuity, and doing it in a half-
thorough compromise, he lost perspicuity anyway.

Appendix D: Linnaeus

Linnaeus (1751) proposed, and Giseke (1792) executed, a table that
seems to give geographical as well as taxonomic information, since Linneaus
expressly compares it to a map, but for Linnaeus it was sheer metaphor. The

table represents degree of resemblance by degree of spatial proximity in two-dimensional space. (Obviously, this presupposes that resemblance, like spatial surface, has just two dimensions.) A. P. de Candolle (1819, p. 231, quoted in Merz 2.243, n.4), says that "Linnaeus was the first to compare the vegetal kingdom to a map; this metaphor, indicated in his book by a single word, has been developed subsequently by Giseke, Batsch, Bernardin de St. Pierre, L'Héritier, Petit-Thouars, etc." And he goes on to elaborate the metaphor: "Classes correspond to parts of the world, families to kingdoms, tribes to provinces, genera to cantons and species to cities or towns". Linnaeus's "single word" is in his *Philosophia botanica*, of 1751, §77, at the beginning of "Fragments, the natural method". The relevant passage, paraphrased by Nordenskiöld (1936, p. 214, cf. p. 438) and quoted with precise reference by Daudin (1926, p. 113, n.2) says: "Nature does not make leaps. All plants show affinity like a territory on a map" (cf. §§ 160, 206, 283). Note that here for Linnaeus, as for Bonnet (according to the interpretation proposed above), continuity (no leaps) does not require unidimensionality (the chain/ladder). Giseke (1792), one of those mentioned by De Candolle as developing Linnaeus's metaphor, offers the 'map' reproduced in Greene p. 140 (with detail on p. 345, n.6 to p. 137). Fifty-four orders (not classes) of plants are represented by circles, the size of each circle reflecting the number of genera in that order. The tenor of Giseke's metaphor (and presumably Linnaeus's) entails that the relative sizes of the circles, and their relative pairwise proximities, are significant, but their orientation is not (as it is in Buffon's table).

Notes

1. See Highet 1949, pp. 275–76 on the latter. The *Port-Royal Grammar* (1660), of the same period, treats modern French as a language just as good as Latin—neither less good nor better. This detached position, by virtually treating the past of the language as nearly irrelevant, was virtually synchronic, a fact remarked on by Saussure (*Cours* 1.3.2).

2. I have at hand the copy of the second edition (April 1863), presented by Lyell to James D. Dana, in which chap. 23 = pp. 454–70. Lyell 1970 adds much information. Huxley's "Darwiniana" (1893–1895, 2: 80–81) discusses Lyell's chap. 23. Tort's (1980) treatment is perceptive.

References

Aarsleff, H. 1967. *The Study of Language in England, 1780–1860*. Princeton: Princeton University Press.

———. 1982. *From Locke to Saussure*. Minneapolis: University of Minnesota Press.

Adler, M. 1940. *Problems for Thomists: The Problem of Species*. New York: Sheed & Ward.

Allen, W. S. 1953. "Relationship in Comparative Linguistics." *Transactions of the Philological Society* 1953: 52–108. Paris.

Becker, K. F. 1827. *Organism der Sprache*. Frankfurt.

Boas, F. 1948. *Race, Language, and Culture*. New York: Macmillan.

Bonnet, C. 1764. *Contemplation de la nature*.

Bopp, F. 1873. *Grammaire comparée des langues indoeuropéennes*, v. 4. Translated by Michel Bréal. Paris: Hachette.

Buffon, George-Louis Leclerc, Comte de. 1755. *Histoire naturelle . . .* , v. 4. Paris: Imprimérie Royale.

————. 1812. *Natural History . . .* , v. 4. Translated by William Smellie; new edition by William Wood. London: Cadell.

Candolle, A. P. de. 1819. *Théorie élémentaire de la botanique*. Second edition.

Cassirer, E. 1950. *The Problem of Knowledge*. New Haven: Yale University Press.

Chambers, R. 1844 ff. *Vestiges of the Natural History of Creation*. London: Churchill.

Cohen, M. R. 1945. *A Preface to Logic*.

Condorcet, A., Marquis de. 1794. *Esquisse d'un tableau historique des progres de l'esprit humaine*. Paris: Agasse.

Cornford, F. M. 1935. *Plato's Theory of Knowledge*. London: Routledge Kegan Paul.

Darmesteter, A. 1886. *La vie des mots*. Paris: Delagrave.

Darwin, C. 1936. *The Origin of Species . . . the Descent of Man*. New York: Modern Library.

Daudin, H. 1926. *De Linné à Jussieu. Methodes de la classification . . .* Paris: Alcan.

Desmond, A. 1982. *Archetypes and Ancestors*. London: Frederick Muller.

Diderichsen, Paul. 1974. "The Foundation of Comparative Linguistics." In Hymes, pp. 277–306.

Donati, Vitaliano. 1758. *Essai sur l'histoire naturelle de la Mer Adriatique*. Traduit de l'italien. La Hague: Pierre de Hondt. (Italian original, 1750.)

Eiseley, L. 1958. *Darwin's Century*. Garden City, N.Y.: Doubleday.

Fischer, R. 1962. "August Schleicher zur Erinnerung." In *Sitzungsberichte der Sächsischen Akademie der Wissenschaften, Philologische-historische Klasse*. 107: 5. Berlin: Akademieverlag.

Ghiselin, M. T. 1969. *The Triumph of the Darwinian Method*. Berkeley and Los Angeles: University of California Press.

Gillispie, C. C. 1951. *Genesis and Geology*. Cambridge, Mass.: Harvard University Press.

————. 1963. "Intellectual Factors in the Background of Analysis by Probabilities." In *Scientific Change* (ed. A. C. Crombie), pp. 431–453. New York: Basic Books.

Giseke, P. D. 1792. *Praelectiones in ordines naturales*. (Cited from Greene 1961, p. 140.)

Greene, J. C. 1961. *The Death of Adam*. New York: Mentor. (First published, 1959.)

Greenough, J. B., and Kittredge, G. L. 1901. *Words and Their Ways in English Speech*. New York: Macmillan.

Gulya, J. ". . . The discovery of Finno-Ugrian." In Hymes, pp. 258–76.

Hage, P., and F. Harary. 1983. *Structural Models in Anthropology*. Cambridge: Cambridge University Press.

Haller, A. von. 1768. *Historia stirpium indigenarum Helvetiae*. Bern: Societas Typographica.

Hanson, N. 1958. *Patterns of Discovery*. Cambridge: Cambridge University Press.

Highet, G. 1949. *The Classical Tradition*. New York and London: Oxford University Press.

Hodgen, M. T. 1952. *Change and History*. New York: Wenner Gren Foundation.

Huxley, T. H. 1897. *Collected Essays*. New York: Appleton.

——. 1901. *Life and Letters of Thomas Huxley*. New York: Appleton.

——. 1898–1902. *The Scientific Memoirs of Thomas Henry Huxley* (ed. M. Foster and E. R. Lankester). 4 vols. London: Macmillan.

Hymes, Dell (ed.) 1974. *Studies in the History of Linguistics*. Bloomington: Indiana University Press.

James, W. 1909. *A Pluralistic Universe*. New York: Longman.

Jespersen, O. 1922. *Language, Its Nature, Development, and Origin*. London: Allen & Unwin.

Jones, Sir William. 1786. *Asiatic Researches* 1: 422.

Kant, Immanuel. 1790. *The Critique of Judgment*. Translated by J. H. Bernard. New York: Hafner, 1951. (German original 1790, 2d ed. 1792; Bernard's translation first published in 1892.)

Kunz, F. L. "The Metric of the Living Orders." In *Integrative Principles of Modern Thought* (ed. Henry Margenau), pp. 291–364. New York: Gordon and Breach.

Lakatos, Imre. 1978. *The Methodology of Scientific Research*. Cambridge: Cambridge University Press.

Lefmann, Solomon. 1870. *August Schleicher, Skizze*. Leipzig: Teubner.

Lovejoy, A. O. 1936. *The Great Chain of Being*. Cambridge, Mass.: Harvard University Press.

Lyell, Sir C. 1863. *The Geological Evidences of the Antiquity of Man*. London: Murray.

——. 1970. *Scientific Journals on the Species Question* (ed. L. G. Wilson). New Haven: Yale University Press.

Madvig, J. N. 1971. *Sprachtheoretische Abhandlungen*. In Auftrage der Gesellschaft für Dänische Sprache und Literatur, herausgegeben von Karsten Friis Johansen. Copenhagen: Munksgaard.

Mencken, J. L. 1936. *The American Language*. 4th ed. New York: Knopf.

——. 1963. *The American Language*. Abridged . . . by Raven I. McDavid, Jr. New York: Knopf.

Merz, J. T. 1904–1912. *A History of European Thought in the Nineteenth Century*. London: Blackwood. (Reprinted New York: Dover, 1965.)

Metcalf, G. 1974. "The Indo-European Hypothesis in the Sixteenth and Seventeenth Centuries." In Hymes, pp. 233–257.

Morpurgo Davies, Anna. 1975. "Language Classification in the Nineteenth Century." In *Current Trends in Linguistics* (ed. T. A. Sebeok). v. 13, pp. 607–716. The Hague: Mouton.

Müller, F. 1876–1888. *Grundriss der Sprachwissenschaft*. Vienna: Holder.

Nordenskiöld, E. 1936. *The History of Biology*. New York: Tudor. (English translation first published in 1928, New York: Knopf.)

Nüsse, H. 1962. *Die Sprachtheorie Friedrich Schlegels*. Heidelberg: Winter.

Ogle, W. 1882. *Aristotle, On the Parts of Animals*. London: K. Paul, Trench.

Ong, W. 1958. *Ramus, Method, and the Decay of Dialogue*. Cambridge, Mass.: Harvard University Press.

Pictet, F. J. 1844–46, 1853–55. *Traité élémentaire de paléontologie ou histoire naturelle des animaux*. Paris: Langlois & Leclerc.

Pott, A. F. 1840. "Indogermanischer Sprachstamm." *Ersch und Grubers Enzyklopaedie* 2.18.

Richards, I. A. 1936. *The Philosophy of Rhetoric*. New York: Oxford University Press.

Ricoeur, P. 1977. *The Rule of Metaphor*. Toronto: University of Toronto Press.

Robins, R. H. 1973. "The history of language classification." In *Current Trends in Linguistics* (ed. T. A. Sebcor), v. 11, pp. 3–41.

Royce, J. 1914. "The Mechanical, the Historical, and the Statistical." Reprinted in *Royce's Logical Essays* (ed. Daniel S. Robinson), pp. 32–62.

Saussure, F. de. 1959. *Course in General Linguistics*. Translated by Wade Baskin. New York: Philosophical Library.

———. 1960. "Souvenirs . . . de sa jeunesse . . ." (ed. R. Godel), *Cahiers Ferdinand de Saussure* 17.15–25.

Schlegel, F. von. 1808. *Über die Sprache und Weisheit der Indier*. Heidelberg: Mohr & Zimmer.

Schleicher, A. [All references are listed in Fischer 1962 (bibliography by J. Dietze), as noted above, p. 46–47.]

Schmidt, J. 1869. "Nachruf (auf Schleicher)." *Zeitschrift für vergleichende Sprachforschung* 18: 315–329.

———. 1890. "Schleicher." *Allgemeine deutsche Biographie* 31: 402–415. Reprinted in *Portraits of Linguists* (ed. T. A. Sebeok) v. 1, pp. 374–395. Bloomington and London: Indiana University Press.

Sneath, P. H. A. 1964. "Mathematics and Classification from Adanson to the present." In *Adanson: The Bicentennial of Michel Adanson's "Famille des plantes."* (ed. H. M. Lawrence) v. 2, pp. 471–497. Pittsburgh: Carnegie Institute of Technology (Hunt Botanical Library).

Spitzer, Leo. 1948. *Essays in Historical Semantics*. New York: S. F. Vanni.

Thompson, Sir. D. W. 1942. *On Growth and Form*. 2d ed. Cambridge: Cambridge University Press.

Tort, P. 1980. *Evolutionnisme et linguistique*. Paris: Vrin.

Trnka, B. 1952. "Zur Erinnerung an August Schleicher." *Zeitschrift für Phonetik* 6: 134–142.

Vorländer, K. 1959 (ed.). *Kant, Kritik der Urteilskraft*. Hamburg: Meiner. (This is an "unveränderter Neudruck der Vorländerschen Ausgabe von 1924," but with different pagination.)

Wells, R. S. 1947. "De Saussure's system of linguistics," *Word* 3.1–31.

Wiseman, Nicholas. 1837. *Twelve Lectures on the Connection between Science and Revealed Religion*. Andover, Massachusetts and New York, New York: Gould and Newman. ("First American from the First London Edition.")

The above essay is an installment in a larger project outlined in a set of lectures on *Comparisons Compared,* delivered at the University of Pennsylvania in 1981 at the invitation of Henry M. Hoenigswald. The next installment, on the concepts of archetype and prototype, is underway.

3

"Organic" and "Organism" in Franz Bopp

ANNA MORPURGO DAVIES

The early nineteenth century is associated in most histories of linguistics with two specific developments: (1) a new conception of language and linguistic work heavily influenced by scientific concepts, taken mostly from the biological sciences, and (2) the creation (if that is the correct word) of comparative and historical grammar. Language in the immediately pre-Romantic, Romantic, and post-Romantic eras is treated as an organism endowed with a life of its own; "organic" is contrasted with "mechanical" and becomes an adjective of praise; terms like growth, decay, weakening, flowering, and branching are in current use in any linguistic discussion; words and roots are said to be related or cognate, and a full set of kinship terms is used to describe the relationship of languages;[1] the comparison between *vergleichende Anatomie* and *vergleichende Sprachwissenschaft,* made famous by Friedrich Schlegel in 1808, becomes commonplace. At the same time, comparative and historical grammars introduce or emphasize a technical and systematic aspect in the subject. It becomes possible to write under the heading *Sprachwissenschaft* highly technical works that are deemed to be as important as the more general essays about the nature and origin of language.[2]

Historiography is now confronted with a series of questions. First, a problem of interpretation: what is the value and meaning of the organic terminology that appears so frequently in any linguistic, and nonlinguistic, discussion? Secondly, what is the connection, if any, between the new organic conception of language and the new "technical" work that characterizes the beginning of comparative and historical linguistics? I leave aside the third important question: what is the origin of this organic explosion in linguistics?

"Organic" and "organism" in nineteenth-century linguistics have often been discussed. The meaning of the words varies according to the authors and often within works of the same author. Organism can be understood in a number of ways. People can refer to the living and developing aspect of the linguistic entity; they can stress the organized nature of the whole, where the emphasis is on the whole as different from the sum of its parts (if this is a meaningful statement); they can concentrate on the functional aspect of the constitutive parts of the organism. Other interpretations are possible.[3] Most of what is important has been said, and I propose to concentrate on my second question: How far does the organic conception of language determine or underlie or simply influence the technical work I mentioned? The question—or its answer—may be of more general import. In a discipline like nineteenth-century linguistics, which oscillates somewhat uneasily between technical work and theoretical discussion, what is the link between theory and practice? Do scholars do what they say they are doing? And do their theories match what they do or what they say they are doing?

The problems I have mentioned are wide-ranging, but they cannot be discussed in a vacuum. In the first instance at least, one must concentrate on a specific case that is taken to be important or representative. Here I shall discuss the work of Franz Bopp (1791–1867), a scholar who rightly or wrongly has been hailed as the founder of comparative and historical linguistics. As the author of the first comparative grammar of the Indo-European languages (1833), Bopp counts as the first and foremost of the "technicians."

Franz Bopp is now a controversial figure. The exalted status attributed to him by his followers has been challenged both on the ground that the so-called conquests of the nineteenth century were an instance of regress rather than progress, and paradoxically on the ground that the greatness of nineteenth-century linguistics is such that it deserves a more profound thinker as an initiator: Friedrich Schlegel or Wilhelm von Humboldt, but not Bopp.[4] Questions of priority or status are irrelevant to my theme, but a correct interpretation of Bopp's views is important, and that too is controversial. As one of the "founders" of comparative linguistics, Bopp automatically qualifies as an exponent of Romanticism like Friedrich Schlegel and Jacob Grimm, but it has been correctly argued that his outlook is much more that of a rationalist. He is interested in understanding the logical structure of language and is closer to Leibniz and the *Grammaire de Port Royal* than to Herder and Friedrich Schlegel (Verburg 1950; Timpanaro 1973). The uncertainty is not dissimilar from that which surrounds Humboldt, but Bopp's simpler personality ought to make it easier to find a label for his views than for those of Humboldt.

We are better informed about Bopp's progress than we used to be.[5] There is little doubt that he came to comparative work because of his interest in oriental languages in general and in Sanskrit in particular, and that at least at first he was heavily influenced by Friedrich Schlegel's *Über die Sprache und Weisheit der Indier* (1808), with its emphasis on a *vergleichende Grammatik* that "would lead to new conclusions about the genealogy of languages in the same way in which *vergleichende Anatomie* had thrown light on *höhere Naturgeschichte*" (Schlegel 1808, p. 25).[6] Friedrich Schlegel used a wealth of organic metaphors in his discussion, but he also distinguished between organic and inorganic languages (i.e., in practice, between the Indo-European languages and all the others) and excluded any link between the two types. Organic languages could only decay from a state of perfection; the other languages improved but never reached organicity. The distinction was based on the nature of the roots. For Schlegel, organic roots were living germs ready to unfold into inflections that indicated the grammatical status of a word. The inorganic languages expressed their grammatical connections "merely" through the juxtaposition of roots. Bopp at first accepted this view and what it presupposed, but already in his first book (1816) he is more critical. The stated purpose of the 1816 work is to discuss the origin of the verbal inflection in the Indo-European languages. The conclusion is that verbal inflection arises in two ways, either by spontaneous alteration or growth of the roots (i.e., organically) or, more frequently, by composition of the verbal stems with other roots, verbal or pronominal. By 1820, Bopp has moved further. Organic inflection is now clearly defined: it refers to reduplication and vocalic alternation in the root, but not to endings and suffixes in general. Still later, Bopp concludes that vocalic alternation too is automatically or mechanically conditioned by the nature of the endings and, at an early stage at least, does not have any grammatical significance. For the mature Bopp, all inflection of the parent language originated in composition. Friedrich Schlegel's theory of the root is dismissed, and so is the distinction between organic and inorganic languages.

So much is known (see Delbrück 1880, pp. 3 ff.; Bréal 1866–1872, 1: xxi ff.; Verburg 1950; Timpanaro 1973), but there are reasons for rehearsing an old story. First, the way in which Bopp proceeds in his attack against Friedrich Schlegel is important for an understanding of his method. His main concern, at the beginning, is to define more clearly the somewhat nebulous concept of organic root. To make it meaningful, he restricts its reference to roots with reduplication and vocalic alternation, and only then does he conclude that Schlegel's typological classification is untenable. It is not supported by the data, because the Indo-European languages have other types of inflection. Also, if the parent language had been "organic" in Schlegel's sense, it could not have functioned as a language. Since the roots were monosyllabic, the possible vocalic alternations would not have been sufficient to indicate all grammatical distinctions.[7] Secondly, it is obvious that we cannot seriously ask what role the concept of organism played in Bopp's work unless we first define his position versus Friedrich Schlegel's theory of the root. *A priori*, we might expect a great number of organic metaphors and organic concepts in the early works, followed by a different attitude after the rejection of Schlegel's views.[8]

In fact, soon after 1816 Bopp stops referring to organic roots and organic languages, yet language is still treated as an organism, and organic metaphors are very frequent. Are we dealing with meaningless words due to the prevailing fashion, or did Bopp really subscribe to a conception of language which is best expressed in organic terms? Bopp's own work was of a very technical nature; if the second hypothesis is correct, what is the connection between Bopp's technical work and his theoretical assumptions?

We can look first at some of the general statements. In 1824, after the final rejection of Schlegel's theory and in the first of the articles that formed the basis for the *Vergleichende Grammatik*, Bopp argues that as time progresses love of euphony gains influence in languages because it is not stopped by a clear feeling for the meaning of individual linguistic elements and because "the branches and ramifications which, as it were, sprouted forth organically in the full life of the language gradually die off and, after they have become a dead mass, can be removed, though the body, which still lives, does not feel this loss."[9] A little later we have the famous passage that appears at the beginning of Bopp's review of Grimm's work (1836; originally published in 1827):

Languages must be taken as organic natural bodies which form themselves according to definite laws, develop carrying in themselves an internal life principle, and gradually die, since they do not understand themselves any longer and shed or mutilate or misuse, i.e., use for purposes for which they were not originally meant, members and forms which initially were significant, but in time have become a mass of largely external nature.[10]

Still later, in the newly published *Vergleichende Grammatik* (1833), we are told that the purpose of the work is to produce "a comparative description of the organism of all languages mentioned in the title summing up all related facts, an investigation of their physical and mechanical laws and of the origin of the forms which indicate grammatical relations."[11]

In the same book, Bopp rejects once again Friedrich Schlegel's classification of languages into organic and inorganic, but finds much sense "in dem Gedanken an eine naturhistorische Classificirung der Sprachen" (1833–1852, p. 112) and proposes to distinguish among (1) languages with monosyllabic roots which are not capable of forming compounds and consequently are "ohne Organismus, ohne Grammatik" (such as Chinese);[12] (2) languages with monosyllabic roots capable of compounding which "acquire their organism, their grammar almost only in this way" ("fast einzig auf diesem Wege ihren Organismus, ihre Grammatik gewinnen") (such as Indo-European and the majority of other languages); and (3) languages with disyllabic roots and three consonants in the root which are the only carriers of lexical meaning (Semitic). The Indo-European languages are deemed to be superior to the Semitic languages, not because of their inflectional pattern but because of the richness of their grammatical complements, of their ability in choosing and using them, and finally because of the "beautiful joining of these complements into a harmonic whole with the appearance of an organic body" ("in der schönen Verknüpfung dieser Anfügungen zu einem harmonischen, das Ansehen eines organischen Körpers tragenden Ganzen." [ibid., p. 113]).

In the technical part of the *Vergleichende Grammatik,* which is the bulk of the work, the style is less lofty and the discussion is highly pragmatic. The method consists of comparing functionally or formally similar segments of different Indo-European languages, concentrating on the formal similarities, and trying to explain away the differences as due to plausible innovations. The first step is to segment the lexical items that are the object of the analysis into roots and inflections. This *Zergliederung* process, which Bopp considers one of his main achievements, is done both on synchronic criteria and on diachronic criteria, though priority is given to the latter. If in a given language synchronic considerations lead to a specific segmentation, this is in the first instance mentioned and accepted; later it can be rejected if at an earlier period of the same language or in related and more archaic forms of another language the *Zergliederung* is different. In ancient Greek, where -*s* is the normal ending of the nominative singular, the genitive corresponding to the nominative *dusmenēs* 'hostile' is *dusmene-os*. If so, we are told, it is natural to assume that in *dusmenēs* the final -*s* is an ending, since it does not recur in other parts of the paradigm (cf., e.g., nom. *phulak-s,* gen. *phulak-os,* where -*s* marks the nominative and -*os* the genitive). On the other hand, two points speak for -*s* as part of the stem: (1) the comparison with the cognate Sanskrit nom. *durmanās,* gen. *durmanas-as,* and (2) the evidence that in Greek -*s*- between vowels drops so that the gen. *dusmene-os* could derive from *dusmenes-os.* A further observation supports the conclusion; -*menēs* is related to the noun *menos* 'anger,' which is a neuter and yet has a final -*s* in the nominative, though -*s* is not an ending used for the neuter nominative (1833–1852, p. 171).[13]

The *Zergliederung* isolates a number of elements (roots, affixes, etc.) that are then compared with each other (though the comparison is partly implicit in the *Zergliederung*) for the purpose of reconstruction. This leads to the

attribution to the parent language of specific roots and inflections. The latter
are analyzed further in the hope of a more complete interpretation. In this way
Indo-European is attributed an -*s* ending of the nominative singular. This is
then compared with the demonstrative pronoun Sanskrit *sa* 'he,' 'this.' The
reasoning is formal rather than semantic: the form *sa* of the pronoun is used
for the masculine and feminine but not for the neuter (Skt. *tat*); similarly, the
-*s* ending occurs in the masculine and feminine but not in the neuter (ibid.,
p. 157).[14]

General statements about organic and organism are not frequent in this
technical part, but the two adjectives *organisch* and *inorganisch* have become
part of the technical jargon: a sound or an ending or a rule is organic when it
belongs to the primitive phase of a language, inorganic when it has been introduced into the language at a later stage.[15]

How should we judge this technical work? There is much that is right in
Kiparsky's (1974) view that the aim is, first, reconstruction of the protoforms
and, secondly, interpretation of the reconstructed morphology in terms of
logical relations. In this sense, as Verburg (1950) pointed out, Bopp is a rationalist, albeit of a special type. Yet after 1816 the emphasis is not so much
on finding in language the logical forms defined by universal grammar as on
recognizing the consistency with which grammatical distinctions are indicated and on identifying the original value of the grammatical affixes.[16] Also,
it is worth stressing that the discussion is limited to a specific language family
and that there is no desire to extend it beyond that point, no pretense of
universality.

I ask again: But why is language treated as an organism? Do the pro-
grammatic declarations quoted above have any purpose in this type of work?
If Bopp is a rationalist, why does he adopt an organicistic conception of language? Should we conclude that the organicistic terminology is verbiage due
to the prevailing trends? Or that we misunderstand the words?

In an interesting, irritating, and at times obscure book, Judith Schlanger
(1971, pp. 125–126) has argued along both these lines:

The term organism may also be a conceptual banality, which is simply equivalent to the
notion of system. This is how Bopp understands it when he declares that his comparative grammar aims at giving a description of the organism of the various languages
that it considers, or that to teach a language means to describe its functioning and its
organism. However, Bopp occasionally does not resist the verbal temptation to introduce a contrast between "understanding the nature and alteration of the simple organism of language" and "attempting to express it through mechanical links", but it is
only a point of detail which remains on the surface of his undertaking. Similarly, to

study the physical or mechanical laws of philology, to treat the physics or physiology of languages—all these formulas aim at expressing the wish for a scientific attitude rather than at indicating a direct analogical link. In Bopp the use of some figurative and facile expressions is simply the result of a slight, unavoidable contamination with the spirit of the times. [17]

No doubt Bopp can use "organism" for system, or, more precisely, for morphological system. In this sense the word is almost synonymous with *Grammatik*. Indeed, in the *Vergleichende Grammatik* (pp. 112–113), quoted above, Bopp glosses the word deliberately in that manner. But why is the gloss provided in one instance and not in the others? Does Bopp's clarity break down at this stage? And is it conceivable that the mature Bopp, who makes fun of at least part of Schlegel's organicistic terminology and complains about its meaninglessness, falls into the same trap? [18]

Let us look again at the aims and method of Bopp's work. A clear statement of intent occurs in the 1836 review of Graff's *Althochdeutscher Sprachschatz:*

[The linguist,] putting together the forms that reciprocally explain each other, finds the most genuine, the most original of them all, and in this way often discovers the reasons for naming an object in a certain manner; thus he clarifies on the one hand the philosophy inherent in language, the aptness of its original assumptions, and on the other hand the regularity and naturalness of its physical composition as well as the simplest constituent elements of its system. [19]

Yet Bopp also explains that in order to reach this result the linguist "establishes for each word as far as possible the regularity of its formation, *and at the same time places side by side with it the course of its development (life)* and describes its appearance in the earlier periods, that is, in the older related languages." [20]

The point made here clearly enough and confirmed by the technical work is that the history of the forms is important not necessarily for its own sake but at least as a means to an end, that is, to reconstruction. In the *Vergleichende Grammatik* the process of reconstruction involves close discussion of the aberrant (i.e., new) forms and of the innovations introduced by each language in the original system. Reconstruction would not be necessary if change had not taken place and languages had not diversified. The point is trivial, but it does need restating. The concrete work is constantly concerned with change and the reasons for, or plausibility of, specific instances of change. [21] Bopp frequently refers to *Wohllautgesetze,* which explain change, and occasionally refers to the *Macht der Analogie,* which accounts for new forms.[22] What is old must be distinguished from what is new. Admittedly, it is what is old, not what is new, that is interesting, but in order to distinguish the two correctly we must understand the why and how of the innovations. Yet, although an understanding of change is so basic for his work and he spends so much time discussing individual instances of linguistic change, Bopp never asks the more general question: why do languages change? In our general picture of Bopp's thought there is a difficulty or inconsistency. We are told that in the early period, when language is in full possession of its vital power (*Lebenskraft*), it also fully understands the meaning and purpose of each word element; in time everything becomes more uniform, the meaning of sounds is less felt and valued so that the sounds may be supplanted and changed (1836, pp. 114–115 [1827]). But what justifies the shift from a period of full understanding to a period of imperfect understanding? Why does a clearheaded scholar, such as Bopp undoubtedly was, not ask the question or even hint at it? Was he too much of a technician to be concerned with general problems? Yet he does ask other general questions, and he was far too close to the beginning of his century and the end of the previous century to be capable of ignoring all general discussion.

In my view, if there is indeed a historiographical problem, it can be solved easily. Bopp did not ask the general question about language change simply because he had implicitly answered it from the start. If language is an organism, not only in the sense that it is a system but also in the biological sense of the term, it must have an inbuilt drive toward change. It develops and decays because this is what organisms do: decay inevitably follows the period of maximum strength. As Cuvier (1800, p. 10) put it, all organisms "croissent par une force intérieure, perissent enfin par ce principe intérieur et par l'effet même de leur vie." There is more: the machine (to take the obvious term of contrast) either functions or breaks, but it does not adapt to different circumstances or replace its broken parts, let alone alter them. Yet it is characteristic of languages to change, to adapt. It may happen that they keep the old rules but that these apply to different linguistic material, or alternatively that they abandon some of the old rules without coming to a complete breakdown of the system. My claim is that Bopp was fully aware of this and, if asked why this was so, would have replied that the phenomenon needed no explanation or at least no explanation at a linguistic level. It was one of the characteristics that language shared with all other organisms.

If I am right, the difficulty or inconsistency mentioned before is removed, or at least pushed one step further from the theory of language to that of organisms, but what evidence do we have for this suggestion? Did Bopp really think in these terms?

I start with some general observations made by Bopp in 1829. They show once again that Bopp was concerned with language change but also that in his mind change or development and organic nature were connected. In a draft reply to Humboldt, who had objected to parts of his theory of the root, Bopp tried to give an account of his method and aims:

So what I say in my grammar about the reasons for, or laws of, linguistic phenomena must always be taken to mean that that is my view and that I have arrived at this conviction through my observation of the development process of language, and that in any instance I may be wrong.[23]

The emphasis is on observation of general processes of change and generalization from there. Bopp continues:

It seems to me necessary so to organize the treatment of a language that it emerges clearly that one (the author) is not (merely) concerned with understanding (reading) the writers of a nation, but also wishes to describe the organism (the development process) of a language for its own sake.[24]

"Development process" (*Entwickelungs[gang]*) here is meant to be added to, or substituted for, "organism" (*Organismus*).

Consider now some examples of technical discussion. In the *Vergleichende Grammatik* (1833–1849, pp. 622ff.) Bopp discussed a presumed oddity of *Altslawisch*. The data were not considered in the second edition of the book because they were faulty, but this need not concern us here.[25] In *Altslawisch*, we are told, the dual of the verb distinguishes a masculine and a feminine in the first and third persons: "We two (males) are" is different from "we two (females) are." Bopp stresses that this was obtained "auf unorganischem Wege, im Abweichung von dem Urtypus unseres Sprachstamms," (in an unorganic way, in contrast with the original type of our language family), which did not distinguish gender in the finite verb. In Bopp's formulation, the endings in question are -*va* (masc.) and -*vje* (fem.) for the first person, and -*ta* (masc.) and -*tje* (fem.) for the third person. Bopp points to the existence of a pronoun *va* 'we two' and postulates the formation of a feminine *vje* 'we two (females)' due to the *Macht der Analogie* with the dual of feminine nouns. The verb, according to Bopp, followed suit, introducing the ending -*vje* to match the inherited -*va*, which was unmarked for gender. In the third person

the original ending -*ta* was formally identical to the pronoun *ta* 'these two.'
On the model of the dual of the nouns, a feminine *tje* 'these two (fem.)' was
created and the verbal ending -*ta* was replaced by two endings, -*ta* and -*tje*,
marked for gender. The conclusion is forcefully stated:

> Yet these feminine verbal endings are remarkable in that they are based on the feeling
> for the grammatical identity of verb and noun[26] and show that the *Sprachgeist* was still
> productively permeated by the close connection which had existed from ancient times
> between the simple pronouns and those linked with the verbal stems.[27]

In other words, language has moved away from the *Urtypus,* but the old
modus operandi of the original organism is still recognizable, though the
forms in question are new.

Elsewhere Bopp argues at length about the origin of the Germanic weak
conjugation and of the typical dental of the preterite (Gothic infinitive *sokjan*,
Ger. *suchen*; Goth. preterite *sokida, * Ger. *suchte*) (1833–1852, pp. 866ff.).
This he identifies with the dental of the verb *thun* 'do' 'make'[28] and finds support for his view in the observation that related classes of Sanskrit verbs (de-
nominatives, causatives, etc.) form their perfect by joining to a verbal noun
the perfect of either *kar-* 'do' 'make' or *as/bhu-* 'be.' In the preface to the
Vierte Abtheilung of the *Vergleichende Grammatik* (1833–1852, 4th Abth.:
vii) it is pointed out that "here too, as in many other things, the apparently
peculiar direction taken by the Germanic languages was, as it were, fore-
shadowed by the ancient sister language of Asia."[29] The general principle is
exactly the same as that formulated for the Slavic forms.

I now piece together the essential elements of another lengthy discussion.
Bopp, as we have seen, argued against Friedrich Schlegel that vocalic alternation in verbal roots was determined by mechanical rules. "Heavy" endings
call for "light" roots and "light" endings call for "heavy" roots; hence, Skt. *e-
mi* 'I go' (with a "heavy" root *e-* < *ai-* and a light ending -*mi*) vs. Skt.
i-mas 'we go' (with a "light" root -*i-* and a "heavy" ending -*mas*).[30] Yet Bopp
is confronted with the problem, which he raises himself, that this distribution
of "heavy" and "light" works for the verbs but not for the nouns. He answers
the objection in different ways. Once he points out that since the rule is not
"original" but arises in the course of the development of the language, there is
no reason why it should apply equally to all forms of the language (1836,
p. 130). Once he stresses the great degree of cohesion of verbal forms in contrast with nominal forms and thus explains the regularity of behavior of verbal
roots. The wording is interesting in itself: "The persons, numbers and tenses
of a verb form a sort of corporate body not only in the paradigms of the grammars but also in reality; they stand to each other in a close family relationship,

which in some way generates a natural feeling of ordering and rank, through which they support each other and, directed by an innate instinct, incorporate the lengthened or shortened root vowel in accordance with the weight of the endings."[31] If we take the two statements together, we must conclude that irregularity is understandable in the application of rules that did not belong to the original organism, and at the same time conclude that the spreading of innovations is determined by the degree of cohesion of the grammatical forms involved. The first point, which leaves one perplexed, makes better sense if taken together with the second and its quasi-structuralistic flavor.

Finally, some light on the way Bopp conceived of language as an organism with an inbuilt capacity for self-destruction and perhaps restructuring comes from his work on non-Indo-European languages. In 1840 he tried to demonstrate that the Malayo-Polynesian languages were derived from Sanskrit. The attempt was unsuccessful in the eyes of both contemporaries and followers, but this does not concern us here. Bopp admitted that he could offer no grammatical proof of kinship, but he argued that the lexical correspondences were too many and too systematic to be due to chance. Linguists, he surmised, found it difficult to conceive of a relationship between Malayo-Polynesian and Sanskrit because they expected the same type of relationship they found between the European languages and either Sanskrit or the parent language. In fact, the European languages "have not experienced any total and radical change, any breakup of their original formation, have not built a new linguistic body from the ruins of a decaying one, but have only undergone individual losses and amputations that do no real damage to the totality of the organism and do not give it an entirely new and strange appearance."[32] The Malayo-Polynesian languages, on the other hand, "have everywhere departed from the grammatical path their mother Sanskrit trod; they have cast off the old garment and put on a new one, or appear, in the South Sea Islands, in full nakedness."[33]

The conclusion is interesting—it is possible for two languages to be related, though one is built on the grammatical ruins of the other. So far the development of a language seems to have been equated with the development of an individual. Here we move somewhat uneasily between the individual and the species. Apparently a language can change its grammatical "type" while remaining in some sense the "same" language. One is reminded of Gabelentz's later theories about a spiral-like development of language (cf. Morpurgo Davies 1975, pp. 630, n.44; 677f., n.124).

How far have we progressed? I suggested earlier that for Bopp the organic metaphors were more than a terminological concession to Romantic fashion. Have I proved my point? It should be clear by now that though a study of language change per se was not the main aim of Bopp's work, it was an indispensable part of it. But if so, some basic questions had to be answered, as I argued earlier. Why do languages change? Why do they change in the way they do? Bopp maintained that if both Germanic and Sanskrit exploited forms of the verbs meaning "to make" in the formation of their perfect, this was not due to change; it was an inherited tendency. Yet the two verbs "to make" were not etymologically related. Does this mean that for Bopp an inherited rule could be preserved *in vacuo,* independently of the linguistic material on which it operated? There is no reason to make of Bopp a deep structure man. The Romantic conception of language as an organism provided a first answer to all these questions and at the same time dispensed the practicing comparativist from any further theoretical effort. Change is built into all organisms: they all grow, decay, and eventually die. All species have a capacity for self-reproduction; different individuals belonging to the same species obey the same instincts, behave in a similar manner, and have similar organs. Why is this so? For the biological organism, Cuvier (from whom I paraphrased most of the statements just made) spoke of a *principe intérieur* that could not be better defined (1800, p. 10). We cannot expect things to be clearer for the linguistic organism. If so, the "technician" is free to leave the problem on one side and get on with his work.

This is all the more true because the biological conception of language also provided a justification for the study of language as such, independently from study of the literary texts. All through Bopp's work this point returns as a leitmotif. In 1815 he declares his intention to change the *Sprachstudium* into "einem philosophischen und historischen Studium," without limiting himself to understand what is written in a language (Lefmann 1891–1897, II, *Anhang* 33*). In a curriculum he explains that having started from a love for oriental literature he found that languages *an und für sich* were no less valuable and important (ibid., 116*). In the 1829 passage quoted above (p. 89) he speaks of his wish to describe the organism of a language (its development) for its own sake. In 1833 he says that his purpose is to treat languages as "Gegenstand und nicht als Mittel der Erkenntniss" in order to offer a *Physik* or *Physiologie* of the languages in question rather than an introduction to their practical use (1833–1852, 1st. Abth.: pp. xiii–xiv).[34] The list could continue, but consider now the implications of the last sentence quoted. If language can be the object of physical or physiological analysis, linguistics can be an autonomous discipline, with a status parallel to that of the other sciences. The basis on which Bopp can argue for this conclusion—or alternatively the price he must pay for it—is the identification of language and biological organism.

Yet while this conception helped with some problems, it was also responsible for other difficulties—concealed to Bopp and his contemporaries but more apparent to their successors. A biological organism is an independent unit. The whole of Bopp's work treats language as wholly independent from the speakers. Their existence is not denied—it hardly could be—but they are entirely ignored. *Sprache, Sprachorganismus,* and *Sprachgeist* all refer to an entity that has a life of its own and obeys its own laws. Other Romantic thinkers equated language with organism, but at the same time tried to find in it an expression of human creativity and looked at it as the perennially developing expression of national identity. This occasionally led them to a paradoxical position similar to that of the literary critic who would enthuse about the organic and self-determined nature of the work of art and then rave about the total freedom enjoyed by the artist in his creative moments.[35] Bopp was too consistent and temperamentally too anti-Romantic to fall into the same trap. Grimm's attempts to equate language development and cultural development found in Bopp a cool listener. At the same time, however, Bopp's analyses of individual instances of change impress the reader as generally impoverished. He ignores the speakers and all problems of language transmission, language diffusion, and so on, and thus deprives himself of a powerful means of explanation.[36]

Moreover, Bopp's concepts of organism, growth, and decay are not consistently applied throughout. In the technical part of his work, we hear a great deal about decay but almost nothing about growth. The assumption is that the separation of the Indo-European languages occurred after the beginning of the linguistic decay of Indo-European; indeed, it was part of it. In Bopp's framework at least, decay seems to lead to loss rather than development, but if this is so, how can new rules and categories be created in a language that is going through a process of decay? Old Slavic, according to Bopp, introduced in the verb a gender distinction that was not present in the *Urtypus.* How is this possible? Organisms have a capacity for self-reproduction, but the new individual is by necessity similar to the old one. If so, how can the Malayo-Polynesian languages derive from Sanskrit but belong to a different type?[37] We may be dealing with an evolutionary theory *in fieri,* but in this case we would expect from Bopp an account of organism that is different from the traditional one or the formulation of an evolutionary principle comparable to Lamarck's inheritance of acquired characters or Darwin's natural selection.

I now turn to a final question and some general conclusions.

How did Bopp come to his concept of organism? Does his reference to *vergleichende Sprach-Anatomie* (e.g., in 1836, p. 83 [1827]) imply a cultural dependence on Cuvier or other scientists? I have no definite answer. Bopp lived in Paris for four years, from 1812 to 1816. He must have heard of the most famous scientists of the period, but we do not know whether he thought about their work or was directly influenced by their views. The important step was the identification of language and organism, yet this step had been taken before Bopp's time, and in Friedrich Schlegel's 1808 book, which he read and studied, he met both the organic conception of language and the reference to *vergleichende Anatomie*.[38] We may look at the way Bopp conceives of an organism as such (not necessarily a linguistic organism). Most of Bopp's statements match those of Georges Cuvier, for example, but similar statements occur in most texts of Romantic and pre-Romantic literature and philosophy. On an abstract level, Cuvier himself did not introduce a new concept of organism. In the first of his *Leçons d'anatomie comparée* (1800, p. 6), Cuvier refers explicitly to Kant, and the whole discussion of the differences between organisms and mechanisms, *corps vivants* and *corps bruts,* is strongly reminiscent of the second part of the *Critique of Judgement*.[39] The point is that Bopp's concept of organism is too vague to be attributed to a particular author or school or even a particular science. Quite simply, it is the concept of organism that was in the air at the time and that anyone could have absorbed without taking the trouble to refer back to the philosophical or biological textbooks where it was explicitly discussed.[40]

    In spite of, or perhaps because of, the vagueness of one of his basic concepts, Bopp is a good example of a phenomenon that keeps recurring in the history of our discipline (and no doubt in that of other disciplines too). The organic views that he accepted were ill-defined but played an important part in his work. I have argued that the identification of language and organism provided him with a justification for the autonomous study of language and an explanation of language change, but it also allowed him to continue his technical work unencumbered by the theoretical worries of his predecessors. Should we then conclude that the theory acted as a sort of conscience-saving device, that it made the concrete work reputable and allowed it to continue but did not interfere with it? To a certain extent this is true, and it serves to explain some of the later developments. Most nineteenth-century linguists saw Bopp as the founding father, not because they agreed with his theories (at some stage these were rejected) or accepted all his results, but because they shared with him a number of common interests and above all because they recognized a fundamental similarity in the techniques which he had used and in those which they were using. Reading Bopp now, almost a century and a half after publication of the *Vergleichende Grammatik,* I have much the same feeling. However, this is not all. Bopp's theory did at times influence his technical work. Occasionally we see that when two interpretations of a phenomenon or a set of forms are possible Bopp chooses the one that better suits his view of the linguistic organism and of the way it works. The new feminine dual forms of the Slavic verb could have been explained analogically (elsewhere Bopp adapts this type of explanation), and in at least one instance Bopp could have thought of analogy with the noun rather than with the pronoun, yet he hints at composition of verb and pronoun. Clearly he is eager to state that his criterion for the formation of the inflected verbs (forming compounds with pronouns) applies to both the early and late stages of the linguistic organism. Elsewhere techniques and theory conflict. Bopp openly admits that comparison leads him to postulate two verbal endings, -mi and -m, for the first person singular of the parent language. Yet he finds it difficult to assume that in *der Jugend-Periode unseres Sprachstamms* there was a double series of endings for one function.[41] This, we are told, could not have happened at a time "when the language organism was still in the full flower of health in all its parts" ("wo der Sprach-Organismus noch in allen seinen Theilen in voller Gesundheit blühte"). Hence the conclusion that -mi was the original ending and that at a later stage -m derived from it through the loss of final -i (1833–1852, p. 633).[42] Why does Bopp in this instance explicitly reject the results of the comparative method? We are reminded of a passage in the *Critique of Judgement* (II, sec. 66) in which Kant defines an "organized product of nature" as "*one in which every part is reciprocally purpose, and means.* In it nothing is vain, without purpose or to be ascribed to a blind mechanism of nature." Kant later continues:

It is an acknowledged fact that the dissecters [*Zergliederer*] of plants and animals, in order to investigate their structure and to find out the reasons, why and for what end such parts, such a disposition and combination of parts, and just such an internal form have been given them, assume as indisputably necessary the maxim that nothing in such a creature is *vain*."[43]

In other words, in dissecting an organism (*Zergliederung* is the word Bopp uses for language segmentation) we must assume that we will not find useless parts.[44] Similarly, Bopp finds it impossible to attribute to a linguistic organism two entirely equivalent endings and is obliged to conclude that one of them arose during the process of decay as a secondary innovation.[45] There is little doubt that in both these instances Bopp's "technical" conclusions are partly determined by his organic view of language. On one occasion he adopts a solution that agrees with his concept of language as a developing organism capable of preserving and transmitting specific rules or modi operandi, on another occasion he modifies some of the conclusions reached through comparison in order to be consistent with his view of the linguistic organism as formed by parts, all of which are functionally necessary.

In summing up, we must summarize, but also look forward at the later developments in the subject. In Bopp's work the organic conception of language, however ill-defined, played an important role. It justified the concentration on language *an und für sich* and provided a theoretical framework (or the simulacrum of one) within which to operate. More concretely, it determined some of the "technical" conclusions. Either the theory or the conclusions (or both) were bound to be influential, and the influence was pervasive and long-lasting. Around the middle of the nineteenth century, Schleicher exploited the organic model far more than Bopp had done (and with much greater awareness), but in so doing he showed the following generations what its limitations were. The neogrammarians reacted against Schleicher and the model, but could not escape the influence of either. The history of nineteenth-century linguistic thought could be rewritten *sub specie organismi,* and I suspect it is the organic conception of Bopp rather than that of Grimm or even Humboldt which determined the later developments.[46]

Notes

1. Not all biological metaphors arose in the nineteenth century. Some of them have a much longer and more illustrious history (see Chapter 1, by W. Keith Percival).

2. In 1923, A. Meillet wrote, "L'expérience montre qu'un fait nouveau bien analysé fait plus pour le développement de la science que dix volumes de principes, même bons" (Meillet 1923–1924, p. 83; cf. also de Mauro 1965, p. 80). ("Experience shows that a new fact, if well analyzed, does more for the development of science than ten volumes of principles, even if good.") It is doubtful that this statement would have been accepted as such in the early nineteenth century, but as early as 1835 it was possible for Bopp (1836), then professor of Sanskrit and *allgemeine Sprachkunde* in the newly founded University of Berlin, to write a lengthy review of a purely technical work like Graff's *Althochdeutscher Sprachschatz* and to hail it as a major contribution to linguistics.

3. See Cassirer 1945; Lepschy 1962–1981 (particularly important); Nüsse 1962, pp. 44ff.; Rensch 1967; Brown 1967, pp. 40ff.; Schlanger 1971; Picardi 1973 and 1977; Timpanaro 1972 (1977) and 1973; and Koerner 1975.

4. For references, see Morpurgo Davies 1975, pp. 610, ff., 623–624.

5. For the literature, see Gipper and Schmitter 1975, pp. 510ff.; 1979, pp. 49–54; and Morpurgo Davies 1975. Add Timpanaro 1973; Antinucci 1975; Paustian 1977; Rousseau 1980; and Koerner 1984. Particularly important are the articles by Verburg 1950, Pätsch 1960, Neumann 1967, and Timpanaro 1973. Lefmann (1891–1897) is still indispensable as a collection of data. Two interesting articles by R. Sternemann (1986 a, b) appeared too late to be considered in this paper.

6. This sentence may be the most quoted statement in the linguistic work of the early nineteenth century. Bopp himself quoted it in translation (Bopp, 1820, p. 15). For Schlegel's influence on Bopp, it is important to see the letters that Bopp wrote from Paris; cf. esp. Lefmann 1891–1897, II, *Anhang* 9*f.

7. See Bopp's letter of March 5, 1820, to Humboldt (Lefmann 1891–1897, Nachtrag 7). See also, e.g., Bopp 1820, pp. 10ff.; 1972, p. 117 (1831, p. 15); 1833–1852, pp. 105ff. See below, p. 90.

8. This distinction is not generally made, but it should be. In what follows, I shall in most cases ignore the 1816 book. No reference to organic, organism, etc., occurs in Bopp's 1820 work, the English article/monograph based on the *Conjugationssystem* (1816). Probably this is due not to a change in Bopp's views between 1816 and 1820 but to the fact that the 1820 work is written in English. As late as the mid-nineteenth century, Ellen Millington, the translator of Friedrich Schlegel's *Aesthetic and Miscellaneous Works* (London, 1849), avoided the words "organic" and "organism" in a linguistic context and preferred to translate the German *Organismus* with "organization" or "structure" or different paraphrases (cf. Schlegel 1977).

9. ". . . und weil die gleichsam in der Lebensfülle der Sprache wie organisch entsprossten Aeste und Verzweigungen nach und nach absterben, und zu einer todten Masse geworden, abgelöst werden können, ohne dass dieser Verlust von dem noch lebenden Körper gefühlt wird" (Bopp 1972, p. 3) [1824, p. 119]).

10. "Die Sprachen sind nämlich als organische Naturkörper anzusehen, die nach bestimmten Gesetzen sich bilden, ein inneres Lebensprinzip in sich tragend sich entwickeln, und nach und nach absterben, indem sie, sich selber nicht mehr begreifend, die ursprünglich bedeutsamen, aber nach und nach zu einer mehr äusserlichen Masse geworden Glieder oder Formen ablegen, oder verstümmeln, oder missbrauchen, d.h. zu Zwecken verwenden, wozu sie ihrem Ursprunge nach nicht geeignet waren." (Bopp 1836, p. 1). This passage, like the Schlegel sentence quoted above, is constantly quoted; cf., e.g., Koerner 1975, p. 735.

11. ". . . eine vergleichende, alles verwandte zusammenfassende Beschreibung des Organismus der auf dem Titel genannten Sprachen, eine Erforschung ihrer physischen und mechanischen Gesetze und des Ursprungs der die grammatischen Verhältnisse bezeichenenden Formen" (Bopp 1833–1852, 1st. Abth., iii).

12. The statement may seem odd in view of Bopp's refusal to distinguish between organic and nonorganic languages. But *Organismus* is here glossed with *Grammatik* and comes close in meaning to our "morphology" or "morphological system." There is no judgment of value similar to that of Schlegel, and there is no reference to an organic development of the roots. (For the background of Bopp's statement, one must probably consider Humboldt's *Lettre à M. Abel Rémusat sur la nature des formes grammaticales en général et sur le génie de la langue chinoise en particulier* (1827); for the identification of grammar and organism, cf. Humboldt 1903–1936, 3: 294ff. [1812]).

13. Contrary to the received opinion, there is no confusion even at this stage between synchrony and diachrony (to use more modern terminology). A striking example of a well-made distinction occurs in Bopp's review of Graff's *Althochdeutscher Sprachschatz* (Bopp 1836, pp. 145–146 [1835]). Bopp knows that the Germanic umlaut arose from an "automatic" alteration of the root vowel which is assimilated to the vowel of the following syllable, but even so he stresses that in modern German the resulting vocalic alternation in certain roots must be attributed a *dynamische Bedeutung,* "because now we do not know that at an earlier stage the -*l* of *Äpfel* [the umlauting plural contrasted with the sing. *Apfel*] was followed by an -*i*- which had an assimilating influence on the preceding *a*-."

14. The point is important. Kiparsky's view (1974, p. 343) that Bopp is a representative of that school of linguistic thought which concentrates on the semantic basis of linguistic form

rather than on the study of the "intricate but perhaps slightly less elusive intrinsic patterning of linguistic forms" seems to be open to challenge. As I point out later, there is a basic continuity between Bopp and the scholars of the second half of the century. This is determined largely by the techniques used in the actual work of reconstruction, and the emphasis is definitely on form rather than on semantics for both Bopp and, for example, the neogrammarians. In a sense Bopp's novelty is here, though this does not prevent him from reaching individual conclusions more on the basis of a faulty semantics or a faulty logic than on formal grounds. I suspect that the general view we have of Bopp has been too much influenced by the beginning of the *Conjugationssystem*, which he wrote when he was twenty-four or twenty-five. For a different approach, see Antinucci 1975.

15. See also Delbrück 1880, p. 18. Pätsch (1960, p. 224) has criticized Delbrück's identification of *organisch* and *ursprünglich* in Bopp but there are many passages in the *Vergleichende Grammatik* where *organisch* is interchangeable with "that which belongs to the parent language," or "*ursprünglich.*" The expression need not have originated with Bopp; see, e.g., Grimm 1822, p. 591: "Ich habe . . . die alth. lautverschiebung als etwas *unorganisches* dargestellt, und freilich ist sie sichtbare abweichung von einem früheren, spurweise noch vorhandenen organismus." ("I have represented the Old High German sound alternation as something *inorganic,* and clearly it is an obvious change from an earlier organism, traces of which still exist"). In a copy of Grimm's *Grammatik* kept in the library of the Taylor Institution in Oxford (10 e.1), an unknown hand has written on the frontispiece: "'Organisch' heisst in meiner Grammatik was der natürlichen Regel der Sprache und ihrer inneren Consequenz gemäss ist, im Gegensatz zu den störenden, unharmonischen Abweichungen.—Jacob Grimm an Halbertsma, 1833." ("In my Grammar 'organic' refers to what is in agreement with the natural rule of the language and its inner functionality, in contrast with the disharmonic, destructive changes.") I have not been able to trace a published reference for this quotation. For the earlier period we must refer to a different but important distinction made by Humboldt in 1812: "Man kann es als einen festen Grundsatz annehmen, dass Alles in einer Sprache auf Analogie beruht, und ihr Bau, bis in seine feinsten Theile hinein, ein organischer Bau ist. Nur wo die Sprachbildung bei einer Nation Störungen erleidet, wo ein Volk Sprachelemente von einem andern entlehnt, oder gezwungen wird, sich einer fremden Sprache ganz oder zum Theil zu bedienen, finden Ausnahmen von dieser Regel statt" (Humboldt 1903–1936, 3: 295). ("One may take it as a firm fact that in a language everything depends on analogy, and that the language structure is an organic structure even in its finer parts. We find exceptions to this rule only where the linguistic formation of a nation undergoes various disturbances, where a people borrows linguistic elements from another, or is compelled to make complete or partial use of a foreign language"). We know that Grimm read the article and approved of it (Tonnelat 1912, pp. 326ff.).

16. Bopp points out more than once that language does not follow strictly logical rules in the choice of its means of expression. He derives the augment, which he takes to be the necessary mark of the past in the parent language, from the negative particle *a-* (the Greek *alpha privativum*) and concedes to his critics that negation of present does not mean past, just as "Negation des Einen ist noch nicht Vielheit (es konnte ja auch Zweiheit, Dreiheit oder gar nichts sein)." ("Negation of the unity is not yet plurality (it could also be duality, triality or even nothing)"). Yet for Bopp the analysis can still be correct because language never expresses something in full but selects for emphasis the most prominent feature or that feature which is taken to be so. An animal that has teeth/tusks is not necessarily an elephant, but in Sanskrit *dantin* 'the toothed/tusked one' means "elephant." In its turn, the word for "tooth" may etymologically mean "the eater," but one could object that not all eaters are teeth; "somit dreht sich die Sprache in einem Kreise von Unvollständigkeiten herum, bezeichnet die Gegenstände unvollständig durch irgend eine

Eigenschaft, die selber unvollständig angedeutet ist" (Bopp, 1833–1852:783–784; cf. also 1820, pp. 26ff.). ("Thus the language moves in a vicious circle of incompleteness; it indicates the objects incompletely, selecting for this purpose one of their qualities, which in its turn is incompletely expressed")

17. "Le terme d'organisme peut, lui aussi, n'être qu'une banalité conceptuelle, qui équivaut simplement à la notion de système. C'est en ce sens que l'entend Bopp, lorsqu'il déclare que sa grammaire comparée se propose de donner une description de l'organisme des différentes langues qu'elle considère, ou qu'enseigner un idiome consiste à en décrire le jeu et l'organisme. Toutefois Bopp ne résiste pas, accidentellement, à la tentation verbale d'opposer "comprendre l'identité et la perturbation de l'organisme simple de la langue" à "en tenter de l'exposer par des liaisons mécaniques": mais chez lui ce n'est qu'un détail à la superficie de son propos. De même, étudier les lois physiques ou mécaniques de la philologie, donner la physique ou physiologie des langues, ces formules cherchent à exprimer l'intention d'une attitude scientifique plutôt qu'elles ne convoient un plein [sic, read 'lien'?] analogique direct. Chez Bopp, le recours à des façons de parler figurées et faciles n'est que la contamination faible et inévitable de l'air du temps." (Schlanger 1971, pp. 125–126)

18. For Bopp's ironical reactions to Schlegel cf., e.g., Bopp 1832, p. 15 "Man mag vorziehen jene Laute [viz., those of suffixes, etc.] gleichsam als die Füsse anzusehen, die einer Wurzel beigegeben oder angewachsen sind, damit sie sich in der Declination darauf bewegen könne; man mag sie auch als geistige Emanationen der Wurzeln ansehen, die, man braucht nicht zu bestimmen wie, aus dem Schoosse der Wurzeln hervorgetreten, und nur einen Schein von Individualität haben, an sich aber Eins mit der Wurzel oder nur ihre organisch entfaltete Blüthe oder Frucht seien. Mir scheint aber die einfachste und durch die Genesis anderer Sprachstämme unterstüzte Erklärung den Vorzug zu verdienen." ("One may prefer to see those sounds, so to speak, as feet, which have grown to a root or have somehow been added to it, so that it can stir itself into a declension; one may also see them as spiritual emanations of the root, which emerge from its lap—no need to say how—and have only an appearance of individuality, but in themselves are at one with root or are merely its organically developed flowers or fruits. It seems to me however that the simplest explanation, which is also supported by the origin of other language families, deserves preference . . .")

19. ". . . durch die Zusammenstellung der sich wechselseitig aufklärenden Formen die echteste, ursprünglichste von allen ermittelt, und hierdurch häufig den Benennungsgrund eines Gegenstandes aufdeckt, und so einerseits die der Sprache inwohnende Philosophie, die Sinnigkeit ihrer Uranschauungen, und andererseits die Regelmässigkeit und Natürlichkeit ihrer physischen Einrichtung, so wie die einfachsten Elemente ihres Ganzen an das Licht zieht" (Bopp, 1836, p. 136).

20. ". . . so weit es möglich ist, einem jeden Worte die Gesetzmässigkeit seiner Bildung nachweist, *ihm gleichsam seinen Lebenslauf zur Seite stellt,* sein Aussehen in früheren Perioden, d.h. in älteren stammverwandten Sprachen beschreibt" (ibid.; emphasis added).

21. In note 45, below, I discuss Bopp's analysis of the identical endings of first and third persons singular perfect in Sanskrit and other languages. Both in 1820 and 1842 Bopp reached the same conclusion: this identity must be due to an innovation. Yet, unlike in 1820, in 1842 he found it necessary to adduce a reason (however implausible this may seem to us nowadays) to justify the innovation.

22. Koerner 1975, p. 734, is right in stressing that the concept of morphological analogy is present in Bopp. This does not, of course, speak against the novelty of the neogrammarians' views on the subject; see Morpurgo Davies 1978.

23. "So ist das was ich in meiner Gr. von Gründen oder Gesetzen der Spracherscheinungen

sage, immer so zu verstehen, dass dies mein Ansicht sey, dass ich durch meine Beobachtung des Entwickelungsgangs der Sprache zu dieser Ueberzeugung gelangt bin, in der mich jedesmal irren kann . . ." (Lefmann 1891–1897, Nachtrag, p. 69).

24. "Es scheint mir notwendig (. . .) die Behandlung einer Sprache so einzurichten, dass man daraus ersieht, dass es einem (dem Verf.) nicht (blos) darum zu thun ist, die Schriftsteller einer Nation zu verstehen (zu lesen), sondern dass man den Organismus einer Sprache (den Entwickelungs[gang]) um seiner selbst willen darstellen will" (Lefmann 1891–1897, Nachtrag, p. 69).

25. The data as we now see them (see Vaillant 1950–1977, 3 1: pp. 15, 27–28) are far more complicated than indicated in the *Vergleichende Grammatik*. In the late texts, Church Slavonic has a third person dual masculine ending *-ta* which is contrasted with a feminine-neuter ending *-tě*. Both forms replace an older *-te* (which was unmarked for gender) and seem to be modeled on the nominal declension. Other Slavic languages (Slovenian, Kashubian, Sorbian) have developed a distinction between the masculine and the feminine both in the pronoun and in some forms of the dual verb. These are late analogical innovations that are difficult to define and to explain.

26. For Bopp, the Indo-European roots were originally distinguished into nominal-verbal roots, which yielded verbs, nouns, adjectives, etc., and pronominal roots, which yielded pronouns, conjunctions, particles, etc. The distinction between nominal and verbal roots was, according to Bopp, secondary and did not belong to the parent language. The point made here is that if the verb can acquire a feminine, though gender distinctions are normally limited to nouns, this shows that there is no fundamental gap between verbs and nouns. Notice that this classification of the roots was something which Bopp had partly inherited from the rationalistic tradition (cf. Verburg 1950, but see also the strictures of Antinucci 1975). He is now trying to find a formal justification for it, and he is doing it in an organicistic framework.

27. "Merkwürdig aber sind jedenfalls diese weiblichen Verbal-Endungen, weil sie auf dem Gefühle der grammatischen Identität des Verbums und Nomens beruhen, und beweisen, dass der Sprachgeist von dem engen Zusammenhang noch lebendig durchdrungen war, der von jeher zwischen den einfachen Pronominen und den mit Verbalstämmen verbundenen bestanden hat" (Bopp 1833–1852, p. 623). According to Bopp, the personal endings were in origin pronominal forms, and the pronominal roots also accounted for most of the grammatical affixes.

28. This explanation replaced an older suggestion by Bopp (e.g., 1816, pp. 118ff.; 1820, pp. 37–38), according to which the dental is to be compared with the one of the participle suffix (Skt. *-ta-*). Both hypotheses are still discussed seriously today.

29. "Es war also hierin, wie in so vielen anderen Dingen, die scheinbar eigenthümliche Richtung, die die Germanischen Sprachen genommen haben, gleichsam schon durch die alte Asiatische Stammschwester vorgezeichnet." (Bopp 1833–1852, 4th. Abth., p. vii).

30. See Bopp 1833–1852, pp. 694ff., and Verburg 1950. Zwirner and Zwirner (1966, pp. 83–84; see also Koerner 1975, p. 734) suggest that in this theory Bopp may have been influenced by contemporary chemistry. As they say, the point is impossible to prove, and the obvious connections between root ablaut and the presence or absence of certain endings may have been sufficient to suggest the thought to Bopp, who also compared the contrast between, e.g., French *je tiens* and *nous tenons*. On the other hand, it is possible that Bopp felt the influence of Karl Heyse's views about the relative "weight" of vowels, views that in their turn depended on the earlier theories of Wolfgang von Kempelen (see Benware 1974, pp. 27ff.). We know that Bopp thought of the alternation as of a "mechanical" law (see Bréal 1866–1872, 1: 1n), and we may ask what determined the use of this adjective. It may be relevant that for the young Bopp, who was still under Schlegel's influence, ablaut or vocalic alternation was the "organic" inflection par excellence. The conclusions he reached later about the alternation in the word of "heavy" root

and "light" ending, or vice versa, meant that he could abandon the "organic" theory of the root and replace it with a nonorganic or, in the jargon of the time, "mechanical" view. In this scenario the influence of chemistry is not necessary. More mysterious is why Bopp, in spite of Humboldt's early hints and the later work by, e.g., Holtzmann, never proposed or accepted the view that vocalic alternation was due to different accentual patterns (cf. Benware 1974, p. 44, and esp. Bréal 1866–1872, 3: xliv–xlvi, with note 1 at p. xlvi).

31. ". . . die Personen, Zahlen, Tempora eines Verbums nicht blos in den Paradigmen der Grammatiken, sondern auch in der Wirklichkeit eine Art von Körperschaft ausmachen, in einem engen Familienverhältniss zu einander stehen, was in ihnen gewissermaassen ein natürliches Ordnungs- und Rang-Gefühl erzeugt, wodurch sie sich wechselseitig unterstützen, und, von einem angeborenen Instinkt geleitet, nach Maassgabe des Gewichtes der Endungen den ausgedehnteren oder eingeengteren Wurzelvocal sich einverleiben" (1836, p. 36 [1827]).

32. ". . . keine totale Umwälzung, keine Auflösung ihres Urbaues erfahren, nicht aus den Trümmern eines zerfallenen Sprachkörpers sich einen neuen gebildet, sondern nur einzelne Verluste und Verstümmelungen erlitten haben, die dem Gesammt-Organismus keinen wesentlichen Abbruch thun, ihm keinen völlig neuen und fremdartigen Anstrich geben" (1840, p. 171 [1972, p. 235]).

33. ". . . sind aus der grammatischen Bahn, worin sich ihre Mutter Sanskrit bewegt hat, überall herausgetreten; sie haben das alte Gewand ausgezogen und sich ein neues angelegt, oder erscheinen, auf den Südsee-Inseln, in völliger Nacktheit" (ibid.).

34. What does *Physik* mean in this context? Grimm's *Wörterbuch* (s.v.) quotes Kant: "Die Wissenschaft der Natur heisst Physik." But is Bopp making a general plea for a comparison with sciences in general, or does he think of the biological sciences in particular? Bréal, who is normally aware of problems of terminology, translates *physique* without further comment (1866–1872, 1: 8); Eastwick (1854, 1: 13) translates *physiology* and omits the previous word. We know that in the same preface "physical laws" refers to grammar rules and especially to phonetic rules. In Bopp 1836, p. 136, quoted above, the *physische Einrichtung* of a language is contrasted with its *inwohnende Philosophie*. I wonder whether in our context *Physik* and *Physiologie* refer to external form and internal organization respectively.

35. For Romantic aesthetic thought and its paradoxes see Abrams 1958, pp. 223–224.

36. We notice the contrast not only with Bredsdorff (1790–1841), who remained virtually unknown, but also with earlier thinkers, e.g., Turgot (I refer to the article "Étymologie" that appeared in 1756 in the *Encyclopédie*). Grimm too, who was much closer to Bopp in the type of work he was doing, strikes us as very different from him in this respect.

37. The question of "type" change is of course controversial. If we assume that Bopp seriously distinguished three linguistic types only, as mentioned above (see p. 84), then we have to acknowledge that in all probability he took Malayo-Polynesian to belong to the same "type" as Sanskrit; they both had monosyllabic roots (a historical concept) and were capable of composition.

38. Qualifications are needed, because once again we come across the basic ambiguity of the term "organism." Cassirer (1953, pp. 153ff.) argues that Herder was the first to formulate the organic concepts that had crucial importance in Romantic thought and that Friedrich Schlegel introduced these concepts into linguistic discussion. Nüsse (1962, p. 44) objected that Humboldt might have priority over Schlegel, and following Arens (1969, p. 171) he referred to an 1895 letter by Humboldt to Schiller where language is called *ein organisches Ganze* (Humboldt 1962, 1: 150). Later than 1808, but closer to Bopp in terminology and outlook, is Humboldt's 1812 article about Basque, with its reference to linguistic *Zergliederung* and its quasi-identification of grammar and organism. Yet there is a considerable difference between those who see in language an autonomous biological organism with its own laws of development and those who simply

stress the systemic, structured, "organic" character of language, or use "organic" as an adjective of praise, or emphasize the organic connection between language on the one hand and thoughts or feelings on the other. Humboldt, Bopp's friend and mentor, cannot belong to the first group if he stresses, as he does repeatedly, that the identification of language and a *Naturkörper* can only be metaphorical (see Picardi 1977, p. 31; Koerner 1975, pp. 741ff.). In the very sentence quoted by Nüsse, he also points out that language is so closely connected with the individuality of the speakers that this *Zusammenhang* cannot be ignored. Where does Schlegel stand with respect to this dichotomy? He does distinguish between organic and nonorganic languages, which at first sight would seem to prevent him from making "organicistic" statements about language in general, and here and there he is certainly concerned with nations and speakers. Yet in 1808 he tends to describe the character of organic languages as if they were autonomous units (Lepschy 1962, p. 185 [1981, p. 54]). Here Bopp is closer to Schlegel than to Humboldt, though he too, like Humboldt, rejects a typology based on the organic/nonorganic distinction. A detailed and factual history of the concept of language as an organism still remains to be written, but even with the data we have, we can see that at some stage a number of scholars oscillated between the various interpretations of organism.

39. Obviously I do not refer to Cuvier's more detailed investigations or to his *principe de corrélation des formes.*

40. Around the end of the eighteenth century and the beginning of the nineteenth, the emphasis on organic and organism in all intellectual circles must have been overpowering. The same terminology recurs in Schelling's *Philosophie der Natur* and, for example, in the works of someone like Cuvier, who disapproved of the obscurity of that brand of German philosophy and of its constant switching between *métaphysique* et *physique* (Flourens 1858, pp. 272ff.); Kant and Goethe are as important in this connection as the early exponents of Romanticism; Savigny constantly refers to the organic nature of language (and law), and so does Humboldt. It looks as if no thinker or scholar, whatever his intellectual background or his theoretical inclinations, could escape the terminological (and ideological) infection. Bopp himself must have been exposed to the contagion in more than one way, but we should not forget the close links between him and his patron K. J. Windischmann, editor of the *Conjugationssystem.* In his turn Windischmann had links with the circles of German Romanticism and for a while was in regular correspondence with Schelling.

41. For Bopp, -*mi* and -*m* are functionally equivalent because in his view the distinction between present and past relies entirely on the absence or presence of the augment.

42. The choice of -*mi* as the original form, which relegated -*m* to the status of an innovation, must have been determined by two motives. First, Bopp's phonological theories allowed him to account far more easily for the loss of a final vowel than for its introduction. Secondly, -*mi* was closer than -*m* to the first person pronoun from which Bopp wanted to derive the ending. We now assume that at a very early stage of Indo-European there was only one ending -*m*; -*mi* was created later, but still in the Indo-European period, adding to -*m* an -*i* element that indicated the *hic et nunc.* There is no difficulty in rejecting (on phonetic and other grounds) Bopp's view, but it is more difficult to justify the decision to attribute to the early phases of the parent language one set of endings rather than two.

43. I have quoted from Bernard's translation (Kant 1931, pp. 280–281). The original text runs: *"Ein organisirtes Product der Natur ist das, in welchem alles Zweck und wechselseitig auch Mittel ist.* Nichts in ihm ist umsonst, zwecklos, oder einem blinden Naturmechanism zuzuschreiben. . . . Dass die Zergliederer der Gewächse und Thiere, um ihre Structur zu erforschen und die Gründe einsehen zu können, warum und zu welchem Ende solche Theile, warum eine solche Lage und Verbindung der Theile und gerade diese innere Form ihnen gegeben worden, jene

Maxime: dass nichts in einem solchen Geschöpf *umsonst* sei, als unumgänglich nothwendig annehmen. . . ." (Kant 1908, pp. 376–77; emphasis in the original).

44. Here the structuralist and functionalist component of Bopp's conception of organism
comes to the fore, but we also see the difference between this approach and that of Saussurean
structuralism.

45. A similar but different problem is discussed in Bopp 1820, pp. 35ff. In Sanskrit the
same ending appears in the first and third person singular and in the second person plural perfect.
Bopp points out that first and third person singular were also identical in Gothic and, in his view,
in Greek (for him the -*a*/-*e* distinction is secondary). Nevertheless he concludes that we should
not reconstruct for the parent language the same form for first and third persons singular perfect:
"However old this rejection of the personal characteristics may be . . . I consider that the omission of the pronominal signs in three different persons, in Sanskrit, was not a defect of the
language in the primitive state." The same conclusion is reached, more explicitly, in Bopp 1833–
1852, p. 857. The comparative evidence, we are told, might lead to the conclusion that the loss of
personal markers had already occurred in the period of unity, before the separation of the Indo-
European languages. Yet this conclusion is not necessary because the presence of the reduplication in the perfect would have provided a natural motivation for the weakening of the endings, so
that the different daughter languages could have followed the impulse independently. No reason is
given for the assumption that there were different endings to start with. If we try to understand
Bopp's motives now, we are bound to be influenced by the view we take of his general background. In rationalistic terms we may remember that for the *Grammaire de Port Royal* a verb is
"un mot qui signifie l'affirmation avec designation de la personne, du nombre, et du temps" (Arnauld and Lancelot 1660, p. 97), ("a word which signifies affirmation while designing the person, number and tense") so that Bopp may feel entitled to look for a different person marker in
the first and third person. On the other hand, we have seen that in an organism each part has a
definite function; what would be the function of a person marker that does not distinguish first and
third person? In fact, we do not even need to think of the *Grammaire de Port Royal* or the organic
view of language. Any traditional grammarian who favored the "analogy" side of the old controversy would have tried to explain away the identity of the endings in question as Bopp did. In this
connection it is interesting to observe that Humboldt's 1812 article, referred to above, combines a
great deal of "organicism" with a number of "analogical" statements.

46. Organicism dies hard. In spite of the earlier attacks by Steinthal, Bréal, Whitney, the
neogrammarians, Saussure, etc., in 1925 the German author of a reputable book on the history of
writing (Jensen 1925, p. 7) found it necessary to quote with total approval a French predecessor
in the same field (Berger 1891, p. ix): "Comme les langues, les écritures sont des organismes
vivants, soumis aux lois de la transformation." ("Scripts, like languages, are living organisms
which obey rules of transformation.") It is ironic that in the same pages Jensen also quotes with
approval two strong enemies of "organicism," Paul and Steinthal (admittedly on different
matters). Less surprising is the dedication to Ernest Renan of Berger's book. In view of Bopp's
attempts to recognize the same modus operandi in different periods of the history of a language, it
is worth remembering that in his Geneva lectures Saussure found it useful to attack this attitude as
the last bulwark of organicism: "En reconnaissant que la prétention de Schleicher de faire de
la langue une chose organique (indépendente de l'esprit humain) était une absurdité, nous con-
tinuons, sans nous en douter, à vouloir faire d'elle une chose organique dans un autre sens en sup-
posant que le génie indo-européen ou le génie sémitique (veille) sans cesse (à) ramener la langue
dans les mêmes voies fatales" (Saussure 1967–1968, 1: 514; cf. also Morpurgo Davies 1975,
p. 678). ("Though we are aware of the absurdity of Schleicher's attempt to make of language
something organic which is independent of human spirit, nevertheless we continue, without real-

izing it, to try to make of it something organic in another sense, since we suppose that the Indo-European genius or the Semitic genius constantly aim at leading language back along the same predestined paths.")

In recent years organic terminology and metaphors have frequently reappeared in linguistic discussion (see, e.g., Chomsky 1975). Yet there is no hint, nor could there reasonably be, of the identification of language and a living organism. De Mauro (1981, p. 239) refers to a statement by Chomsky (1975, p. 11) according to which "the idea of regarding the growth of language as analogous to the development of bodily organism is thus quite natural and plausible." It is with some relief that one discovers that "bodily organism" is a misprint for "a bodily organ." The question concerns, of course, the problem of language acquisition by the child. In comparing language to something that grows on the human organism, like wings on a bird, Chomsky is relatively close to the early Romantics and above all to Humboldt.

References

Abrams, M. H. 1958. *The Mirror and the Lamp: Romantic Theory and the Critical Tradition.* 2d ed. New York: Norton.

Antinucci, F. 1975. "I presupposti teorici della linguistica di Franz Bopp." In *Teoria e storia degli studi linguistici.* Atti del settimo convegno internazionale di studi, Roma 2–3 giugno 1973, a cura di U. Vignuzzi, G. Ruggieri, R. Simone, 1: 153–174. Roma: Bulzoni.

Arens, H. 1961. *Sprachwissenschaft. Der Gang ihrer Entwicklung von der Antike bis zur Gegenwart.* Zweite Auflage. Freiburg: Alber., 1st ed., 1955.

Arnauld, A., and Lancelot, C. 1660. *Grammaire générale et raisonnée.* Facsimile ed. Menston: Scolar Press 1967.

Benware, W. A. 1974. *The Study of Indo-European Vocalism in the 19th Century, from the Beginnings to Whitney and Scherer.* Amsterdam: Benjamins.

Berger, P. 1891. *Histoire de l'écriture dans l'antiquité.* Paris: Imprimerie Nationale.

Bopp, F. 1816. *Ueber das Conjugationssystem der Sanskritsprache in Vergleichung mit jenem der griechischen, lateinischen, persischen und germanischen Sprache.* Herausgegeben von K. J. Windischmann. Frankfurt a.m.: Andrea.

——. 1820. *Analytical comparison of the Sanskrit, Greek, Latin and Teutonic languages, showing the original identity of their grammatical structure.* Annals of Oriental Literature 1: 1–64. Reprint, Amsterdam: Benjamins, 1974.

——. 1825. *Vergleichende Zergliederung des Sanskrits und der mit ihm verwandten Sprachen. Erste Abhandlung. Von den Wurzeln und Pronominen erster und zweiter Person.* Abhandlungen der Königlichen Akademie der Wissenschaften, pp. 117–48. Reprinted in Bopp 1972, pp. 1–32.

——. 1832. *Vergleichende Zergliederung des Sanskrits und der mit ihm verwandten Sprachen. Fünfte Abhandlung. Ueber den Einfluss der Pronomina auf die Wortbildung.* Abhandlung der Königlichen Akademie der Wissenschaften, pp. 1–28. Reprinted in Bopp 1972, pp. 103–130.

——. 1833–1852. *Vergleichende Grammatik des Sanskrit, Zend, Griechischen, Lateinischen, Litthauischen, [Altslawischen,] Gothischen und Deutschen.* 6 Abtheilungen. Berlin: Dummler.

——. 1836. *Vocalismus, oder sprachvergleichende Kritiken über J. Grimm's Deutsche Grammatik und Graff's Althochdeutschen Sprachschatz mit Begründung einer neuen Theorie des Ablauts.* Berlin: Nicolai.

————. 1840. *Über die Verwandtschaft der malayisch-polynesischen Sprachen mit den indisch-europäischen.* Abhandlungen der Königlichen Akademie der Wissenschaften, pp. 171–246. Reprinted in Bopp 1972, pp. 235–310.

————. 1972. *Kleine Schriften zur vergleichenden Sprachwissenschaft.* Gesammelte Berliner Akademieabhandlungen 1824–1854. Leipzig: Zentralantiquariat.

Bréal, M. 1866–1872. *Grammaire comparée des langues indo-européennes* par François Bopp. Traduite sur la deuxième édition et précedée d'une introduction par M. Michel Bréal. 4 vols. Paris: Imprimerie Impériale/Nationale.

Brown, R. L. 1967. *Wilhelm von Humboldt's Conception of Linguistic Relativity.* The Hague and Paris: Mouton.

Cassirer, E. A. 1945. "Structuralism in Modern Linguistics." *Word* 1: 99–120.

————. 1953. *The Philosophy of Symbolic Forms I: Language.* Translated by R. Manheim. New Haven and London: Yale University Press. Original German edition: Berlin, 1923.

Chomsky, N. 1975. *Reflections on Language.* New York: Pantheon.

Cuvier, G. 1800. An viii. *Leçons d'anatomie comparée.* Paris: Baudouin.

Delbrück, B. 1880. *Einleitung in das Sprachstudium. Ein Beitrag zur Methodik der vergleichenden Sprachforschung.* Leipzig: Breitkopf & Hartel.

De Mauro, T. 1965. *Introduzione alla semantica.* Bari: Laterza.

————. 1981. Position paper in "Tavola Rotonda di QdS su 'Il concetto di natura umana in Chomsky.'" *Quaderni di Semantica* (Bologna) 2: 235–280.

Eastwick, E. B. 1854. *A Comparative Grammar of the Sanscrit, Zend, Greek, Latin, Lithuanian, Gothic, German and Sclavonic languages* by Professor F. Bopp. Trans. from German by E. B. Eastwick. 2d ed. Vol. 1. London: James Madden.

Flourens, P. 1858. *Histoire des travaux de Georges Cuvier.* 3d ed. Paris: Garnier.

Gipper, H., and Schmitter, P. 1975. "Sprachwissenschaft und Sprachphilosophie im Zeitalter der Romantik." In Sebeok 1975, pp. 481–606.

————. 1979. *Sprachwissenschaft und Sprachphilosophie im Zeitalter der Romantik.* Tübingen: Narr.

Grimm, J. 1822. *Deutsche Grammatik.* Erster Theil. Zweite Ausgabe. Göttingen: Dietrich.

Humboldt, W. von. 1812. "Ankündigung einer Schrift über die Vaskische Sprache und Nation, nebst Angabe des Gesichtspunctes und Inhalt derselben. *Deutsches Museum* 2: 485–502. Reprinted in Humboldt 1903–1936, 3: 288–299.

————. 1903–1936. *Gesammelte Schriften.* Herausgegeben von der Königlich-Preussischen Akademie der Wissenschaften. 17 vols. Berlin: Behr.

————. 1962. *Der Briefwechsel zwischen Friedrich Schiller und Wilhelm von Humboldt.* Herausgegeben von Siegfried Seidel. 2 vols. Berlin: Aufbau.

Jensen, H. 1925. *Geschichte der Schrift.* Hannover: Lafaire.

Kant, I. 1908. *Kant's gesammelte Schriften.* Herausgegeben von der Königlich Preussischen Akademie der Wissenschaften, Band 5. Berlin: Reimer.

————. 1931. *Kant's Critique of Judgement.* Translated with introduction and notes by J. H. Bernard. 2d ed. London: Macmillan.

Kiparsky, P. 1974. "From Palaeogrammarians to Neogrammarians." In *Studies in the History of Linguistics,* ed. D. Hymes, pp. 31–43. Bloomington: Indiana University Press.

Koerner, E. F. K. 1975. "European Structuralism: Early Beginnings." in Sebeok 1975, pp. 717–827.

————. 1984. "Professor Dr. h.c. Franz Bopp." *Aschaffenburger Jahrbuch für Geschichte Landeskunde und Kunst des Untermaingebiets* 8: 1–7.

Lefmann, S. 1891–1897. *Franz Bopp, sein Leben und seine Wissenschaft.* 2 vols. and
    Nachtrag. Berlin: Reimer.
Lepschy, G. C. 1962–1981. "Osservazioni sul termine struttura." *Annali della Scuola
    Normale di Pisa* 31: 173–197. Reprinted in G. C. Lepschy, *Mutamenti di pros-
    pettiva nella linguistica* (Bologna, 1981), pp. 37–71.
Marouzeau, J. 1943. *Lexique de la terminologie linguistique.* 2d ed. Paris: Geuthner.
Meillet, A. 1923–1924. Review of *Linguistique et dialectologie romanes,* by G. Mil-
    lardet. *Bulletin de la Société de Linguistique de Paris* 24 (2d fasc.): 80–84.
Morpurgo Davies, A. 1975. "Language Classification in the Nineteenth Century." In
    Sebeok 1975, pp. 607–716.
————. 1978. "Analogy, Segmentation, and the Early Neogrammarians." *Transac-
    tions of the Philological Society,* 1978: 36–60.
Neumann, G. 1967. "Franz Bopp—1816." In *Indogermanische Sprachwissenschaft
    1816 und 1966,* pp. 5–20. Innsbrucker Beiträge zur Kulturwissenschaft, Son-
    derheft 24. Innsbruck: Leopold-Franzens-Universität.
Nüsse, H. 1962. *Die Sprachtheorie Friedrich Schlegels.* Heidelberg: Winter.
Pätsch, G. 1960. "Franz Bopp und die historische-vergleichende Sprachwissen-
    schaft." In *Forschen und Wirken. Festschrift zur 150 Jahr-Feier der Humboldt-
    Universität zu Berlin 1810–1960,* 1: 211–228. Berlin: Deutscher Verlag der
    Wissenschaften.
Paustian, P. R. 1977. "Bopp and Nineteenth-Century Distrust of Indian Grammatical
    Tradition." *Indogermanische Forschungen* 82: 39–49.
Picardi, E. 1973. "Organismo linguistico e organismo vivente." *Lingua e stile* 8:
    61–82.
————. 1977. "Some Problems of Classification in Linguistics and Biology, 1800–
    1830." *Historiographia Linguistica* 4: 31–57.
Rensch, K. H. 1967. "Organismus—System—Struktur in der Sprachwissenschaft."
    Phonetica 16: 71–84.
Rousseau, J. 1980. "Flexion et racine: trois étapes de leur constitution J. C. Adelung,
    F. Schlegel, F. Bopp." In *Progress in Linguistic Historiography,* ed. E. F. K.
    Koerner, pp. 235–247. Amsterdam: Benjamins.
Saussure, F. de. 1967–1968. *Cours de linguistique générale.* Edition critique par R.
    Engler, I. Wiesbaden: Harrassowitz.
Schlanger, J. 1971. *Les métaphores de l'organisme.* Paris: Vrin.
Schlegel, F. 1808. *Über die Sprache und Weisheit der Indier.* Heidelberg: Mohr &
    Zimmer. Reprinted in Schlegel 1977.
————. 1977. *Über die Sprache und Weisheit der Indier.* New edition with an intro-
    ductory article by Sebastiano Timpanaro, translated from the Italian by J. P.
    Maher, prepared by E. F. K. Koerner. Amsterdam: Benjamins.
Sebeok, T. 1975. *Current Trends in Linguistics,* ed. T. A. Sebeok. Vol. 13: *Histo-
    riography of Linguistics,* 2 vols. The Hague and Paris: Mouton.
Sternemann, R. 1984a. "Franz Bopps Beitrag zur Entwicklung der vergleichenden
    Sprachwissenschaft." *Zeitschrift für Germanistik* 5: 144–158.
————. 1984b. *Franz Bopp und die vergleichende indoeuropäische Sprachwissen-
    schaft.* Innsbrucker Beiträge zur Sprachwissenschaft. Vorträge und Kleinere
    Schriften 33. Innsbruck: Institut für Sprachwissenschaft.
Timpanaro, S. 1972. "Friedrich Schlegel e gli inizi della linguistica indoeuropea in
    Germania." *Critica storica* 9: 72–105. Translated as "Friedrich Schlegel and
    the Beginnings of Indo-European Linguistics in Germany" in Schlegel 1977,
    pp. xi–xxxviii.

————. 1973. "Il contrasto tra i fratelli Schlegel e Franz Bopp sulla struttura e la genesi delle lingue indoeuropee." *Critica storica* 10: 1–38.

Tonnelat, E. 1912. *Les frères Grimm. Leur oeuvre de jeunesse.* Paris: Colin.

Vaillant, A. 1950–1977. *Grammaire comparée des langues slaves.* 5 vols. Lyon and Paris: IAC; Paris: Klincksieck.

Verburg, P. A. 1950. "The Background to the Linguistic Conception of Franz Bopp." *Lingua* 2: 438–468. Reprinted in *Portraits of Linguists,* ed. T. Sebeok, vol. 1, Bloomington: Indiana University Press, 1966, pp. 221–250.

Zwirner, E., and Zwirner, K. 1966. *Grundfragen der Phonometrie.* Zweite Auflage. Erster Teil. Basel and New York: Karger.

4

On Schleicher and Trees

KONRAD KOERNER

Understanding

We are still far from fully understanding the general intellectual climate of the nineteenth century, although it is clear that the twentieth century has built on the nineteenth to a considerable extent. In linguistic historiography, the subject of "influence" has remained an elusive one, frequently used differently by different authors. Perhaps even the distinction between "direct" and "indirect" influence, or what may be ascribed to the general "climate of opinion" of a given period, has not proved sufficiently useful in the debate. But it is clear that fledgling fields of study, such as linguistics in the nineteenth century, have always tended to borrow from fields (not always adjacent or related) that in the eyes of the informed public had already attained a high degree of scientific elaboration, with regard to both their methods and their findings. It is now widely accepted that nineteenth-century historical-comparative linguistics was profoundly inspired by the natural sciences, in particular botany, comparative anatomy, and pre-Darwinian evolutionary biology (see Koerner 1980). This does not exclude the influence, acknowledged or not, of philology on linguistics. In fact, one would expect there to be such influence, since linguistics had its most immediate source in the much older field of philology, with many nineteenth-century linguists not only having received their training with well-established philologists, but also continuing to contribute to philology even after their major concerns had shifted to linguistics.

In my view, August Schleicher was the most important nineteenth-century linguist, so it is justified to pay so much attention to his background and general interests as well as to his scholarly achievements (Hoenigswald 1963,

1974, 1975). It appears to me now that differing views on the background sources of Schleicher's linguistic theories can to a considerable extent be reconciled.

Misunderstanding

In a monograph-length study on what I termed the "early beginnings of structuralism" in linguistics, completed in early 1972, I argued against Hoenigswald's (1963, p. 5) opinion that it was Schleicher's philological training under Friedrich Ritschl (1806–1876), which included the establishment of stemmata depicting the relationship of manuscripts and their possible descent from a common source, that led Schleicher to the family-tree idea, arguing instead that this suggestion must remain an open question "since Schleicher never referred to the analog" (Koerner 1975, p. 755 n. 62). I maintained this position as late as 1981, when I expanded on my views of the "Schleicherian paradigm" of the early 1870s (see Koerner 1982, p. 31, n. 24). I admit that this sounds like bad methodology on my part, and in contradiction to statements of method made elsewhere (e.g., Koerner 1976), according to which the educational background of a scholar, including family tradition, should be taken into account when writing on the history of linguistics. Indeed, although Schleicher did his doctorate under Ritschl on a philological subject which may have included stemmatics, I am surprised that his biographer Joachim Dietze (1966, p. 18) affirms: "Die Keime zu seiner [i.e., Schleicher's] späteren den Naturwissenschaften ähnlichen Forschungsmethode hat Ritschl gelegt" (the seeds for his [i.e., Schleicher's] later research method, which was similar to that of the natural sciences, had been sowed by Ritschl).

Dietze did not expatiate on this observation, and in personal correspondence he stated that he deduced this from a statement made in the dictionary entry of 1890 by Schleicher's pupil Johannes Schmidt (1843–1901). Dietze's reference to Schleicher's *later* research method as inspired by the natural sciences is somewhat misleading; it implies that Schleicher had developed and subsequently abandoned an earlier method. Besides, it is tantalizing to imagine that Ritschl the philologist should have instilled in Schleicher an orientation derived from the natural sciences, since this would combine two what I believe to be distinct sources of Schleicher's inspiration. If I did not embrace Hoenigswald's suggestion concerning the probable source of Schleicher's family trees in historical linguistics, it was not to deny that Schleicher was familiar with the philological methods developed by Karl Lachmann (1793–1851) and others from the late 1810s onward, but that the establishment of stemmata was a much less suggestive source for Schleicher's familiar concrete trees as depicted from 1853 on (see Priestly 1975, pp. 301, 302).

More important—and this appears to be the main reason for a misunderstanding—Schleicher does not appear to have used tree diagrams of the inverted kind typically used by Lachmann, Ritschl, and others (see Timpanaro 1971, pp. 46, 48, and elsewhere), with the supposed source on top and the later manuscripts, copies, and so on, branching downwards (see the various trees in Schleicher's *Die deutsche Sprache* [1860], which are quite different even from Darwin's diagram in *Origin of Species,* of which he took note only in 1863). Furthermore, Schleicher used his tree model to depict relationships between languages—with respect to each other and to the parent language, not between individual forms, akin to what we refer to as the triangulation technique of the comparative method. As far as I can see, the use of an inverted tree diagram in his method of reconstruction (and Schleicher was a pioneer in historical reconstruction in linguistics) was implicit but nowhere stated. If Hoenigswald refers to the method of reconstruction when he speaks of the analogue between this technique and that of the philologists, he may well be right, but if this model is supposed to be the immediate source of Schleicher's genealogical trees, I still believe I should contradict him.

Intellectual Biography

Schleicher's father was a medical doctor, and it was in his early youth that Schleicher took a strong interest in nature, in particular botany. Various biographical accounts point to this interest of Schleicher's and his spectacular successes in plant breeding. In addition, Schleicher himself refers to this interest in his writings, in particular his well-known "open letter" of 1863 to the apostle of Darwinism in Germany, Ernst Haeckel (1843–1919), in which he also shows that he is quite familiar with Darwin's predecessors. Indeed, since the publication of J. P. Maher's paper (1966), we should desist from calling Schleicher a Darwinian. He never became one, but remained a pre-Darwinian evolutionist all his life (see Koerner 1982, pp. 5–8, for details).

In a 1967 paper, "Zur Geschichte der Stammbaum-Darstellungen," Georg Uschmann, who at the time did not yet know of Schleicher's family trees, traced early statements concerning "arbres généalogiques" back to the second half of the eighteenth century, though it appears that the earliest depictions of tree diagrams in a scientific work promoting the idea of evolution are to be found in Lamarck's voluminous *Philosophie zoologique* of 1809 (see Uschmann 1967, p. 13). As early as 1766, French botanist Antoine Nicolas Duchesne (1747–1827) used the term *arbre généalogique* in his *Histoire*

naturelle des fraisiers (Paris: Didot le Jeune). In the same year the German
naturalist Peter Simon Pallas (1741–1811), who is best known to linguists as
editor of the Russian enterprise with the Latin title *Linguarum totius orbis
vocabularia comparativa* (St. Petersburg, 1787–1789), suggested in his
Elenchos zoophytorum (see Uschmann 1967, p. 11):

Unter allen übrigen bildlichen Vorstellungen des Systems der organischen Körper
würde es aber wohl die beste sein, wenn man an einen Baum dächte, welcher gleich
von der Wurzel an einen doppelten, aus den allereinfachsten Pflanzen und Tieren
bestehenden, also einen tierischen und vegetabilischen, aber doch verschiedentlich
aneinanderkommenden Stamm hätte. (It would probably be the best of all figurative
conceptions of the system of organic bodies, if one was to think of a tree which had,
already from the root onwards, a double, albeit by different means connected, stem
consisting of the simplest plants and animals, respectively.)

There is no reason to believe that Schleicher was familiar with this work, but
his own statement of 1853, introducing his genealogical tree of the Indo-
European language family, appears to echo Pallas' suggestion: "Diese Annah-
men, logisch folgend aus den Ergebnissen der bisherigen Forschung, lassen
sich am besten unter dem Bilde eines sich verästelnden Baumes anschaulich
machen" (these assumptions [i.e., of an IE language family], deduced logically from the results of previous research, can best be depicted by the image
of a branching tree) (Schleicher 1853, p. 787). Given Schleicher's long-
standing penchant for the natural sciences and his early struggle for a science
of language independent of and in strong contrast to philology (see Schleicher
1850, pp. 1–5), it is more likely that Schleicher allowed himself to be guided
by principles developed by the natural scientists much more than by those due
to the philologists of his time. This does not exclude a much more subtle "influence" on him by the latter.

References

Dietze, J. 1966. *August Schleicher als Slawist: Sein Leben und sein Werk in der Sicht
    der Indogermanistik.* Berlin: Akademie-Verlag.
Hoenigswald, H. M. 1963. "On the History of the Comparative Method." *Anthro-
    pological Linguistics* 5, 1: 1–11.
———. 1974. "Fallacies in the History of Linguistics: Notes on the Appraisal of the
    Nineteenth Century." In *Studies in the History of Linguistics,* ed. Dell Hymes,
    pp. 346–360. Bloomington and London: Indiana University Press.
———. 1975. "Schleicher's Tree and Its Trunk." In *Ut Videam: Contributions to an*

Understanding of Linguistics: For Pieter A. Verburg . . . , ed. Werner Abraham
et al., pp. 157–160. Lisse and Holland: P. de Ridder Press.
Koerner, E. F. K. 1975. "European Structuralism: Early Beginnings." In *Current
Trends in Linguistics,* ed. Thomas A. Sebeok, vol. 13: *Historiography of Linguistics,* pp. 717–827. The Hague: Mouton.
———. 1976. "Towards a Historiography of Linguistics: 19th and 20th Century Paradigms." In *History of Linguistic Thought and Contemporary Linguistics,* ed.
Herman Parret, pp. 685–718. Berlin/New York: W. de Gruyter.
———. 1980. "Pilot and Parasite Disciplines in the History of Linguistic Science."
Folia Linguistica Historica 1: 213–224.
———. 1982. "The Schleicherian Paradigm in Linguistics." *General Linguistics* 22:
1–39.
Maher, J. P. 1966. "More on the History of the Comparative Method: The Tradition of
Darwinism in August Schleicher's Work." *Anthropological Linguistics* 8, 3:
1–11.
Priestly, T. M. S. 1975. "Schleicher, Čelakovský, and the Family-Tree Diagram: A
Puzzle in the History of Linguistics." *Historiographia Linguistica* 2: 299–333.
Schleicher, A. 1850. *Die Sprachen Europas in systematischer Uebersicht.* Bonn: H. B.
König. New edition, with introduction by Konrad Koerner. Amsterdam: J. Ben-
jamins, 1983.
———. 1853. "Die ersten Spaltungen des indogermanischen Urvolkes." *Allgemeine
Monatsschrift für Wissenschaft und Literatur,* 1853, pp. 786–787.
———. 1860. *Die deutsche Sprache.* Stuttgart: J. G. Cotta.
———. 1863. *Die Darwinsche Theorie und die Sprachwissenschaft: Offenes Send-
schreiben an Herrn Dr. Ernst Häckel . . . .* Weimar: H. Böhlau.
Schmidt, J. 1890. "Schleicher, August." *Allgemeine deutsche Biographie* 31: 402–
416. Berlin: Duncker & Humblot.
Timpanaro, S. 1963. *La genesi del metodo del Lachmann.* Florence: F. Le Monnier.
———. 1971. *Die Entstehung der Lachmannschen Methode.* Translated into German
by Dieter Irmer. Hamburg: H. Buske.
Uschmann, G. 1967. "Zur Geschichte der Stammbaum-Darstellungen." In *Gesammelte Vorträge über moderne Probleme der Abstammungslehre,* ed. Manfred
Gersch, vol. 2, pp. 9–30. Jena: Universität Jena.

5

A Legal Point

BOYD H. DAVIS

In trying to determine the relationships between Indo-European languages, linguists in the nineteenth century moved from the metaphor of "family" to the model of a branching tree. The biological nature of the model seems obvious; other disciplines offered interpretations of a tree model, whch reinforced its appropriation. This brief note is not comparative evidence, but it is evidence for adoption of a metaphor for the comparative method. It looks briefly at the genetic model as presented by nineteenth-century historians of Roman law. A substantial portion of Roman law was based on cognatic and agnatic affiliation, a genetic model based on the power (potestas) of the father which was used, among other things, to determine the division of property. The power of the metaphor is seen in two examples. The first is a controversy over the nature of linguistics between two would-be fathers of linguistic thought. Here the division of property was a division of scholarly thought, and kinship often determined which scholars or notables took whose side. Such a division of property—here intellectual spheres of inquiry—is not uncommon. A second example, drawn from the literature of early England, examines the legitimization of relationships and the granting of authority to Beowulf through the particular speech acts taking place within the poem: the power of the father is transferred and the division of property is effected.

In a discussion of the move away from a scale or chain of being, Thomas Goudge (1973, p. 197) comments: "By the time of Charles Darwin, another metaphor had come to the fore, namely that of a 'great tree' whose twigs, branches, boughs, etc., represent respectively species, genera, families, etc. of living things, ramifying in a complex, irregular way from a single trunk, or from two main trunks at the base." He refers to this tree by a common metaphor: the "tree of life." This visual concept is so well internalized in so many

Western belief systems that it is an almost commonplace emblem, a visible archetype accessible to the learned and illiterate alike. Woodcuts, altar paintings, illustrations in emblem books, engraved title pages—all expand the details given to and furnished by the tree. An iconographer of Christian art might trace through medieval and Renaissance artifacts a movement from the tree with Adam, Eve, and the serpent to the tree alone, as a catalog of vices, virtues, or in some cases kingly lineages. The iconographer seeks context, traces suggestions, and studies associations. Some of the same methods must be used in an exploration of the potency of the tree model in its appropriation and retention by nineteenth-century linguists.

Hoenigswald (1963) demonstrated that early linguists were influenced in their choice of a model for charting language relationships by the stemmatics practiced by contemporaneous philologists. Stemmatics worked from a kind of branching-tree model to display genetic affiliations among texts.

The ways in which a language was studied were interdependent with the ways in which its text was established and interpreted. Early linguists also felt additional influence from the courses in the "new" approaches to history and jurisprudence evolving at German universities. It is frequently noted that Jacob Grimm was a devoted pupil of the historian Fredrich Carl von Savigny; few have explored the possible connections between the models chosen by earlier linguists and historians. This link is not a direct one.

Jacob Grimm studied law and even collected laws for his *Deutsches Recht,* which he dedicated to Savigny, but he was apparently not interested in devising any laws. Grimm's "law" governing affiliations or changes of shared innovations in sounds, was not called a law for nearly half a century. However, one can study legal principles in practice or principles for the study of law without being compelled either to invent or to implement law. To seek direct, causal connection here is a blind end, but an examination of associations and relationships suggests explanations and provides missing pieces to a puzzle that does not neatly bifurcate.

G. P. Gooch has noted that Savigny's lectures on the history of jurisprudence were clear, brilliant, and crowded. Their focus was the continuity of Roman law during the medieval period. For Savigny, law operated very like language in that it was based on custom and principle, its organic growth a response to the needs of its peoples. Says Gooch: "The jurists were no more the authors of law than the grammarians of language; they only developed what the folk life created." Part of this product remained customary, while other portions were turned into "laws" (1959, pp. 44–45). Whether discussed by Savigny or remade in the later seminars of Leopold Ranke, Roman law in its various forms of survival was a focal point for many nineteenth-century historians and legal scholars. Georg Iggers reminds us that in 1825, at the University of Berlin, Hegel taught philosophy, Karl Friedrich Eichhorn and Savigny taught law, and Ranke taught history. They attracted students from America as well as the Continent. Iggers notes that Ranke's methods for studying the "documents of modern history were basically, of course, similar to those applied by the great classical philologists of the late eighteenth and early nineteenth centuries" (Iggers and von Moltke 1973, pp. xxiv–xxv).

By mid-century, much of the clerisy, to use Samuel Taylor Coleridge's term for the aggregate of scientists, litterateurs, and theologians, actively pursued their studies either under these German scholars or under their followers. Regardless of the door through which they entered, these scholar-gypsies were exposed to content that had as its bedrock an emphasis on genetic affiliation, progress, and decay.

The fifth edition (third American edition) of Sir Henry Summer Maine's *Ancient Law* of 1875 provides a fine illustration of the extension of the historical and philological models. The first (London) edition was published in 1861, the year that first saw Max Müller's *Lectures on the Science of Language* and in which Maine was invited, like many British scholars before him, to take a high post in the Indian government. The fifth edition carries an introduction by Theodore W. Dwight, Columbia College's professor of municipal law (the very existence of this professorship is resonant of the study of *lex in urbe* promoted by the German historians and jurists). The introduction is really a state-of-the-art overview of Maine's writings with its focus on this Maine's first in a series of major studies. B. E. Lippincott notes that *Ancient Law* "created a prodigious stir, exercising in its own field an influence comparable to that of Darwin's *Origin of Species* (1938, p. 171).

Having studied under the leading German scholars, Maine learned not only "to apply the doctrine of evolution to human institutions" but also that law and social institutions were formed by custom (Lippincott 1938, pp. 170–171). Often credited for establishing the "natural history of law," Maine's work continues to arouse admiration. Thorner (1980, p. 258) praises Maine for setting the course of historical jurisprudence and for giving "a demonstration of the comparative methods which, for sheer brilliance . . . has not been surpassed." India was his example of a society still embodying social and legal institutions similar to earlier states in Western societies. *Ancient Law* traced laws back through successive codifications to their prototypes in custom, finding the patriarchal family to be the core.

When we turn to Dwight's introduction, we read one of the primary modes of disseminating Maine's methods and ideas throughout the American clerisy during the latter half of the century. Dwight went to great pains to make clear Maine's central thesis as exemplified by Roman law, the *patria potestas,* or the authority of the patriarch over the persons and property of descendants.

Kinship was of fundamental importance. Dwight's table (Figure 5.1) is designed to explain the two fundamental relationships in Roman law: cognate and agnate. Cognates are all blood relations tracing descent from one legitimate marriage. Agnates trace descent exclusively through males. Dwight's note on the Mexican and Austrian discussions of succession are a reminder to the twentieth-century reader that the Napoleonic code and discussions of European succession were keenly examined in the America of the 1870s (Maine 1875, pp. xxxiii–xxxv).

Though there were other methodologies for history and jurisprudence in the nineteenth century, the subject matter of much of ancient law concerned kinship and property. The methodology evidenced by Maine's extension of the German historians was comparative in the philological sense. Linguists in the first half of the century, then, felt forces from several disciplines that influenced their choice of the branching-tree model. And its retention? By the appearance of later editions of Maine's comparative study of legal institutions, new theories about the representation of language relationships were surfacing. Earlier principles of language had come to be perceived as "laws"; debates over new approaches show how entrenched the earlier model had become. Kinship, this time among scholars themselves, played a part in these debates. For example, twentieth-century readers of *The Double Helix* will not be surprised to learn that family connections entered into scholarly polemic.

The polemic between the American scholar William Whitney and the German-born Oxford don Max Müller, which extended over the larger part of the latter nineteenth century, is interesting in and for itself. Their debate

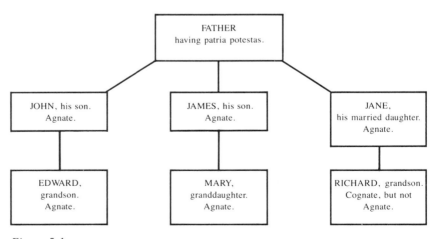

Figure 5.1

touched on many subjects, the most important of which were the nature and origin of language, the classification of linguistics as either a natural science or a historical science, and the philosophy and methodology proper to linguistic investigation. Their quarrel was carried by a wide range of publications. Issues were raised not only in reviews and articles designed for learned societies, but also in public lectures, reviews, letters, and articles intended for the educated public on both sides of the ocean. It is probable that several of the points in their conflict, such as the notion of phonetic decay, the correctness of etymologies, and the preeminence of Sanskrit for historical reconstruction, were not fully grasped by the nonspecialist. However, the general tone of transatlantic debate, its basis in "things Oriental," and its involvement with the Darwin family invited an outspoken partisanship, that continued long after the death of both participants.

While both Whitney and Müller had remained fairly balanced in the assessment each had offered of the other's work, the fat was in the fire when the family connections began to intervene. For instance, family was no small matter in English intellectual life. At a time when only forty percent of the male population was allowed to vote, somebody's attendance at one or another college of Cambridge or Oxford, or his family relationships outside school, were no trivial matters. Affiliations were sources of legitimization.

Charles Darwin's son George was grandson, on his mother's side, to the famous potter Josiah Wedgewood and cousin to Hensleigh Wedgewood, one of the original members of the Philological Society (*Dictionary of National Biography,* 20, pp. 1051–1052). Hensleigh had already disagreed publicly with Müller on the origin of language. When, in 1873, Müller's lectures at the Royal Institution, which disapproved of any portion of Darwinian theory that would affect the possibility of the divine origin of language, were reported in *Fraser's Magazine,* George had had enough. He published an article entitled "Professor Whitney on the Origin of Language" (Darwin 1874). The next ten years involved both Whitney and Müller in replies fueled by this article. In 1892, Whitney was still combating what he felt to be Müller's lack of coherence. Largely unnoticed, except by linguists, were the fine points that each scholar offered the other for the common task of establishing precise models to explain language relationships. Instead, as with the discussion by Whitney of August Schleicher's organic theories, writers from several camps entered their own opinions on the role of evolution or the duty of Darwin.

In their polemic, as with similar debates over competing methods in the twentieth century regardless of field, neither Whitney's nor Müller's adherents considered whether two models might, after all, be needed in some cases. There are times when more than one model is needed if a particular problem is to be elucidated. For example, language may be considered both conventional and capable of codification. A precedent, again, a legal one, exists: the legitimization of a speaker before the king and court in early Germanic society. The *Beowulf,* as well as other heroic Anglo-Saxon verse, contains a series of speeches introduced by the form "X maðelode." Speech-act analysis of this work shows that it introduces certain discourse patterns. The topics are knowledge and wisdom; the speakers are heroic, noble warriors of either sex. The rhetoric of these speeches calls forth responses that have been documented to some extent by both Tacitus and Julius Caesar. Without going into detail (see Davis 1980) on the Indo-European ramifications of the root involved, one can say that these speeches appear in specific contexts and elicit certain responses. A speech introduced by this verb can be challenged, interrupted, debated. The Unferth episode in *Beowulf,* and the speech of the seaguard to Beowulf and his men, are two examples of how challenge can be presented or forestalled. The speech is always in the presence of the king and counselors, or in their tacit presence, as sanctioned by custom and delegated responsibility. Once heard by the assembly, even if only one is present, the content is assumed to be experienced as true by all hearers. The speaker's legitimacy, veracity, and leadership is formally accepted. The legitimacy is built as much on affiliation with known persons as on the cogency of the argument. The audience for the heroic poem is part of the process of legitimization, becoming part of the assembly by the skill of the poet.

What biologists call the "reticulate" model may pick up the additional evidence and establish the context that the genetic/stemmatic/cladistic model cannot hypothesize, since the relationships needed for the speaker go beyond the cognatic/agnatic and move into other spheres of reference. Germanic tribal chieftains, nineteenth-century linguists, and twentieth-century scientists share similar problems based on how they are willing to apportion the known universe of discourse about knowledge and to assign legitimacy to the proponent. Thoughts change. We cannot always be sure which of us is Unferth.

References

Darwin, G. 1874. "Professor Whitney on the Origin of Language." *Contemporary Review,* November 1874, pp. 894–904.
Davis, B. 1980. "Maðelode: Telling as Knowing in *Beowulf.*" Manuscript circulated to Old English Discussion Circle, South Atlantic Modern Language Association.
Gooch, G. P. 1959. *History and Historians in the Nineteenth Century.* Paperback with new introduction from first edition (1913). Boston: Beacon Press.
Goudge, T. 1973. "Evolutionism." *Dictionary of the History of Ideas.* ed. P. Wiener. New York: Scribner.

Hoenigswald, H. 1963. "On the History of the Comparative Method." *Anthropological Linguistics* 5: 1–11.

Lippincott, B. E. 1938. *Victorian Critics of Democracy.* London & Minneapolis: University of Minnesota Press.

Maine, H. S. 1875. *Ancient Law.* Introduction by T. Dwight. 3d American ed. from 5th London ed. New York: Holt.

Muller, F. M. 1861–1864. *Lectures on the Science of Language.* London: Longmann, Green.

Ranke, L. 1973. *The Theory and Practice of History.* Introduction and ed. by G. Iggers and K. von Moltke. New York: Bobbs-Merrill.

Thorner, D. 1980. "The Comparative Method of Sir Henry Maine." In *The Shaping of Modern India,* pp. 257–272. New Delhi: Allied Publishers.

"Wedgewood, Hensleigh," *Dictionary of National Biography* xx: 1051–1052. Oxford: Oxford University Press, 1927.

Whitney, W. D. 1892. *Max Müller and the Science of Languages.* New York: Appleton.

6
Haeckel's Variations on Darwin

JANE M. OPPENHEIMER

The subject of the symposium for which this chapter was originally prepared was "Biological Metaphor Outside of Biology." Why, then, the two biologists named in my title? Ernst Heinrich Haeckel, who dealt with evolutionary trees, took the ideas of Charles Darwin as his leaping-off points. It was Haeckel as a popularizer who transmitted the biological ideas of his times, especially those of Darwin, to the general reader. He started out as a zoologist, but his strictly scientific publications were of much less influence than his popular writings, which drew wide public attention to Darwin's ideas, especially in Germany and in English-speaking countries.

In developing my theme, I shall lean heavily on pictorial examples of evolutionary trees or nontrees chosen from several of Haeckel's works. An earlier authoritative treatment of Haeckel's position in the history of development of the concept of family trees was presented by Georg Uschmann in 1967[1] from a somewhat different point of view. Haeckel's ideas about evolution were published first in 1866 in a more or less technical work, *Generelle Morphologie*,[2] which was not well received, for reasons soon to be specified. He later exploited its ideas in two popular books: *Natürliche Schöpfungsgeschichte* (1868)[3] and *Anthropogenie* (1874).[4]

Haeckel first read the *Origin of Species* in the summer of 1860 in Bronn's German translation.[5] He first lectured on Darwin, speaking "Für die Entwicklungstheorie Darwins," in 1863 at a meeting of the *Deutsche Naturforscher.*[6] In November 1861 he wrote his bride that he was deep in Darwin.[7] According to excerpts from his manuscripts now in Jena, he was in 1864 at page 442 of Bronn's translation. His notes were terminated only in November 1864, that is, after he delivered the 1863 lecture.[8]

In 1865, according to his autobiography, he lectured publicly on Darwin in Jena,[9] where he was a professor. Notes taken by several students on the 1865 lectures formed the basis of *Generelle Morphologie,* written in 1865–1866.[10] This was addressed to professional zoologists, but they were very critical of it. It appeared only in the one original edition and has never been translated. Haeckel had planned an English edition and wrote to ask T. H. Huxley's help with it,[11] but the plan was abandoned, presumably because of the heavy criticism heaped on the German edition. Haeckel said later that it may have been unsuccessful because of its "voluminous and unpopular style of treatment, and its too extensive Greek terminology."[12] These were strong faults, but also facts and relationships were greatly exaggerated and over-systematized.

Haeckel lectured again on Darwin in 1867–1868. These lectures became the basis of the popular book, *Natürliche Schöpfungsgeschichte,*[13] a popularization of the ideas developed in the more technical *Generelle Morphologie.* In fact, *Generelle Morphologie* included all Haeckel's key ideas. They were simply elaborated on in the popular books on which he mainly concentrated hereafter. In Volume 2 of *Generelle Morphologie,* Haeckel had included what he called a *Stammbaum* of man, a genealogical tree of man. In 1871–1872, in the introduction to his own *Descent of Man,* Darwin wrote that Haeckel had "recently" (1868, with a second edition in 1870) published the *Natürliche Schöpfungsgeschichte,* which contained a full discussion of the genealogy of man. "If this work had appeared before my [Darwin's] essay had been written, I should probably have never completed it. Almost all the conclusions at which I have arrived I find confirmed by this naturalist, *whose knowledge on many points is much fuller than mine.*"[14] I would dispute Darwin on the final point: Haeckel's "knowledge" was sometimes supplied by imagination rather than by the scientific process. We know that Darwin did not read German; the first English edition of *Natürliche Schöpfungsgeschichte* appeared only in 1876, five years after the publication of *The Descent of Man.* Perhaps Huxley, who had a fine knowledge of German, helped him to understand the German text.

The *Natürliche Schöpfungsgeschichte* passed through twelve editions in Germany and at least twenty-five translations, many of them in English under the title *The History of Creation.* As we have already pointed out, it was followed in 1874 by *Anthropogenie,* which passed through seven German editions and many English editions under the title *The Evolution of Man.* While these were the principal books in which Haeckel presented to the lay public the ideas first worked out in *Generelle Morphologie,* he continued to write popular books. Some of the titles of English editions are: *The Riddle of the Universe; Wonders of Life; God-Nature; Crystal-Souls; Monistic Building*

Stones; and *The Dagger of Darkness.* These selected titles give an inkling of Haeckel's style of thought. References to publications by Haeckel cover almost seventeen pages in the *National Union Catalogue of Pre-1956 Imprints.*

Haeckel's lectures were also popular. His roll books are available; over his whole teaching career at Jena (1861–1909) he lectured to 6,436 students. The maximum number of students attending his lectures in a single semester was 163, in 1865–1866, the year before any of his books appeared. These were probably the lectures on which the *Generelle Morphologie* was based. (He did publish another book, on anatomy, in 1866.) Seven times, more than one hundred students enrolled in his lectures for a semester. To set these numbers in context, the population of Jena in those years was about 6,000; 400–500 were students.[15]

Generelle Morphologie applied the theory of evolution to the whole organic kingdom, including man. It laid heavy emphasis on the role of embryology in establishing what evolutionary relationships had been. This idea was hardly original with Haeckel; J. H. F. Kohlbrugge[16] counted up seventy-two earlier usages of the idea up to 1866. This is perhaps worthy of comment because of subsequent historians' emphasis on the importance of Haeckel's so-called biogenetic law, which states that ontogeny recapitulates phylogeny, but it is irrelevant to our argument here, where our principal interest is the nature of Haeckel's pictorial illustrations.

Darwin, in the *Origin of Species* at the end of the chapter on natural selection, wrote a lyrical and eloquent passage comparing the affinities of all the beings in the same class to a great tree:

I believe this simile largely speaks the truth. The green and budding twigs may represent existing species; and those produced during each former year may represent the long succession of extinct species. At each period of growth all the growing twigs have tried to branch out on all sides, and to overtop and kill the surrounding twigs and branches, in the same manner as species and groups of species have tried to overmaster other species in the great battle for life. The limbs divided into great branches, and these into lesser and lesser branches, were themselves once, when the tree was small, budding twigs; and this connection of the former and present buds by ramifying branches may well represent the classification of all extinct and living species in groups subordinate to groups. Of the many twigs which flourished when the tree was a mere bush, only two or three, now grown into great branches, yet survive and bear all the other branches; so with the species which lived during long-past geological periods, very few now have living and modified descendants. From the first growth of the tree, many a limb and branch has decayed and dropped off; and these lost branches of various sizes may represent those whole orders, families, and genera which have now no living representatives, and which are known to us only from having been found in a fossil state. As we here and there see a thin straggling branch springing from a fork low down in a tree, and which by some chance has been favoured and is still alive on its summit, so we occasionally see an animal like the Ornithorhynchus or Lepidosiren, which in some small degree connects by its affinities two large branches of life, and which has apparently been saved from fatal competition by having inhabited a protected station. As buds give rise by growth to fresh buds, and these, if vigorous, branch out and overtop on all sides many a feebler branch, so by generation I believe it has been with the great Tree of Life, which fills with its dead and broken branches the crust of the earth, and covers the surfaces with its ever branching and beautiful ramifications.[17]

But Darwin's diagram in the same chapter, reproduced here as Figure 6.1, was highly schematic. It consisted of straight lines only, ascending from a number of points of origin at the bottom of the page. Some of the lines branched, others did not.[18] Haeckel included in *Generelle Morphologie* eight plates, illustrations of supposed genealogical origins, which he entitled *Stammbäume*. These looked something like branching trees, but not exactly. They were flat, with only occasional slight indications of three-dimensional perspective. Long parallel lines, sometimes broken, apparently drawn freehand, might have seemed to represent bark. For an example, Figure 6.2 (Plate VIII of *Generelle Morphologie*) reproduces Haeckel's pictorial representation of an evolutionary tree of mammals, including man. Haeckel was an artist of sorts, and had he wished to draw a less diagrammatic tree, his abilities would have permitted him to.

The text of *Natürliche Schöpfungsgeschichte* presented the same kind of material that *Generelle Morphologie* did, not only more popularly but also, if possible, more dogmatically. This book included a number of synoptic tables (organizational charts we might call them technically; Haeckel called them *Systematische Übersichten*), which represented pedigrees not only of organisms but also of human races. The final table, for instance, included what Haeckel considered "twelve species of men and their 36 races."[19] Some of these tables were made up of outlines consisting of proper nouns only; others were accompanied on the pages opposite by pedigrees (*Stammbäume* again), which included, in addition to the proper nouns, straight lines representing descent and sometimes some brackets. In his text he discussed very briefly the evolution of language, referring to August Schleicher and Fritz Müller,[20] but he presented no schema or diagram to represent it. That would follow in *Anthropogenie*.

The pictorial figures in *Natürliche Schöpfungsgeschichte* represented not evolutionary trees but stylized bushes—nothing having any counterpart in nature. Figure 6.3 of this essay (Plate V in *Natürliche Schöpfungsgeschichte*, 3d ed.), representing the pedigree of the plant kingdom; Figure 6.4 (Plate VI in *Natürliche Schöpfungsgeschichte*, 3rd ed.), representing the monophyletic origin of six animal phyla; and Figure 6.5 (Plate XIV of *Natürliche Schöpfungsgeschichte*, 3rd ed.), representing the pedigree of vertebrates, are examples of some of the fanciful bushlike growths Haeckel invented to illustrate *Natürliche Schöpfungsgeschichte*. The one in Figure 6.4 suggests leaves as well as stems, but it is difficult to know what it really signifies because it resembles no natural growth known on this planet. My figures, from the third edition, are altered from the versions of the same plates found in the first edition. It would be another study revelatory of Haeckel's thought patterns to trace the evolution and change of Haeckel's illustrations, and indeed of his text, in successive editions of a single book.

Anthropogenie, which first appeared in 1874, concentrated on embryological development as explanatory of evolutionary development. The full title of the book was *Anthropogenie, oder Entwickelungsgeschichte des Menschen*. Here the German language served Haeckel well, since it uses the word *Entwickelung* (*Entwicklung* in modern spelling) as the equivalent of the English concepts of both embryonic development and evolution). In this book Haeckel exploited to the fullest his concept that ontogeny, the development of the individual, recapitulates that of the stock to which the individual belongs. The hypothesis that phylogeny is the mechanical cause of ontogeny was elevated by the argument in this book to the status of supposedly immutable law.

It was in *Anthropogenie* that he expanded his treatment of the evolutionary derivation of languages to a long passage of over six pages,[21] and here too he added a synoptic table illustrating a pedigree of Germanic languages. Gone, however, here are the flat tree-like figures of *Generelle Morphologie* and the fanciful bushlike excrescences of *Natürliche Schöpfungsgeschichte*. There is but a single pictorial representation of a family tree, this time of the pedigree of man (Figure 6.6, Plate XII of *Anthropogenie*). The tree is not imaginary. It is shown in a drawing, in full three-dimensional perspective, of a real tree, so real that it can be classified taxonomically. It is a *Quercus robur*, an unmistakable European oak, gnarls and all. Haeckel liked oaks. Elsewhere in *Anthropogenie*, in another connection, he wrote, "When we enter an oak forest we express our awe before the venerable thousand-year-old trees in rapturous words."[22]

Thus Haeckel transformed what was diagrammatic in Darwin to a striking visual image for which Darwin had used only words. Was Haeckel recalling Darwin's verbal description of the tree in *The Origin of Species* when he drew his great oak? Probably not. It is more likely that one kind of drawing

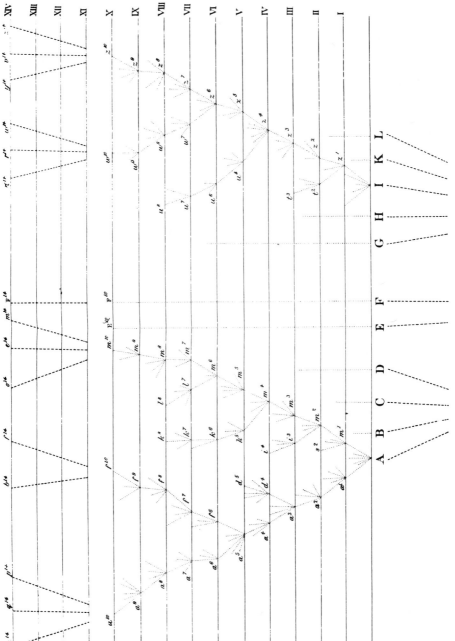

Figure 6.1. Diagram illustrating the chapter on Natural Selection in the first edition of Darwin's *On the Origin of Species.*

Figure 6.2. An evolutionary "tree" of mammals, including man (Plate VIII of *Generelle Morphologie*). There is little suggestion of three-dimensional perspective.

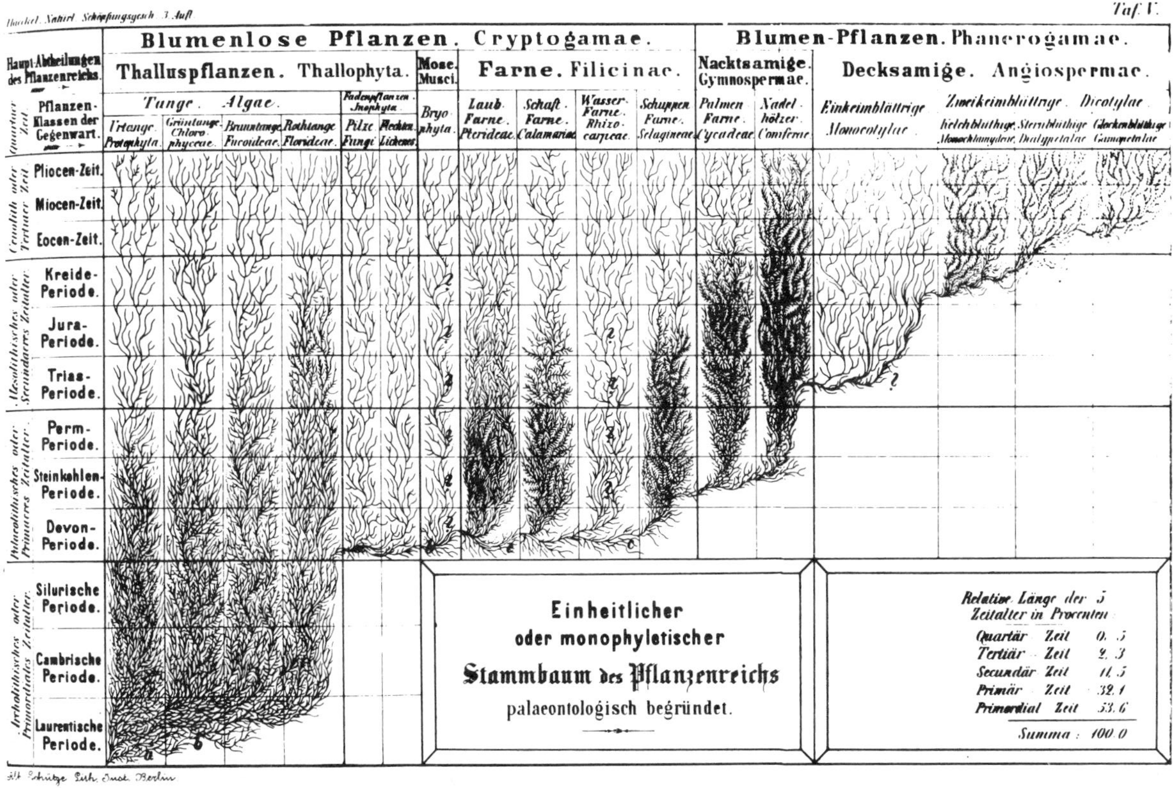

Figure 6.3. Monophyletic pedigree of the plant kingdom, based on paleontology (Plate V of *Natürliche Schöpfungsgeschichte*, 3d ed.).

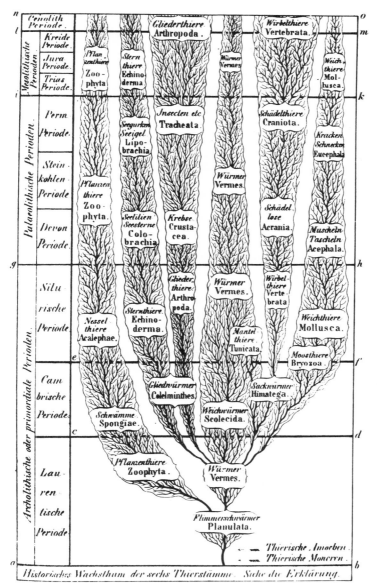

Figure 6.4. The monophyletic origin of six animal phyla, based on paleontology (Plate VI of *Natürliche Schöpfungsgeschichte*, 3d ed.).

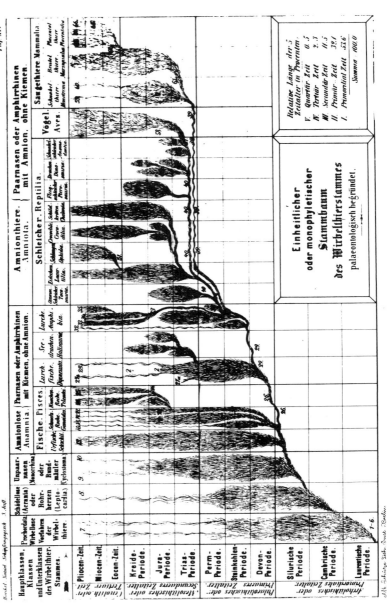

Figure 6.5. A monophyletic pedigree of vertebrates, based on paleontology (Plate XIV of *Natürliche Schöpfungsgeschichte*, 3d ed.).

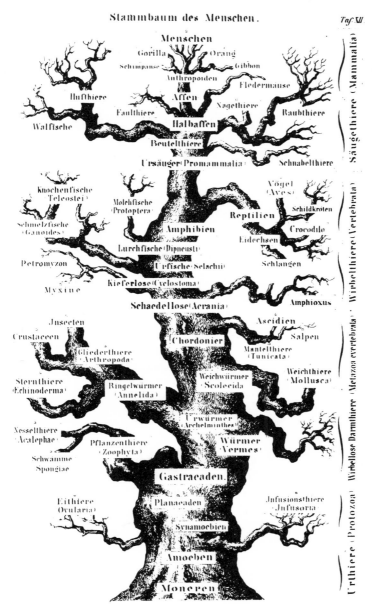

Figure 6.6. The family tree of man, an unmistakable European oak (Plate XII of *Anthropogenie*).

simply led to another in his illustrations, inducing him to transform what was diagrammatic in Darwin to a realistic and vivid picture. Most likely, he was simply attempting to improve on himself, or on nature.

It was a failing of Haeckel as a would-be scientist that his hand as an artist altered what he saw with what should have been the eye of a more accurate beholder. He was more than once, often justifiably, accused of scientific falsification, by Wilhelm His[23] and by many others. For only two examples, in *Anthropogenie* he drew the developing brain of a fish as curved, because that of reptiles, birds, and mammals is bent.[24] But the vesicles of a fish brain always form in a straight line. He drew the embryonic membranes of man[25] as including a small sac-like allantois, an embryonic organ characteristic of and larger in reptiles, birds, and some nonhuman mammals. The human embryo has no sac-like allantois at all. Only its narrow solid stalk remains to conduct the umbilical blood vessels between embryo and placenta. Examples could be multiplied significantly.

Haeckel set a high value on his illustrations, and unrealistic though many of them may have been, there is no doubt that they helped to encourage the general acceptance of Darwin's ideas, including those concerning evolutionary trees.

Notes

1. Georg Uschmann, "Zur Geschichte der Stammbaum-Darstellungen," in *Ges. Vorträge über moderne Probleme der Abstammungslehre,* ed. Manfred Gersch (Jena: Friedrich-Schiller-Universität, 1967), 2: 9–30.

2. Ernst Heinrich Haeckel, *Generelle Morphologie,* 2 vols. (Berlin: Georg Reimer, 1866).

3. Ernst Heinrich Haeckel, *Natürliche Schöpfungsgeschichte* (Berlin: Georg Reimer, 1868). The third improved edition published by Reimer in 1872 has been used for the preparation of this chapter.

4. Ernst Heinrich Haeckel, *Anthropogenie, oder Entwickelungsgeschichte des Menschen* (Leipzig: Wilhelm Engelmann, 1874).

5. Georg Uschmann, *Geschichte der Zoologie und der zoologischen Anstalten in Jena* (Jena: Gustav Fischer, 1959), p. 41 n. 194.

6. Ernst Heinrich Haeckel, "Eine autobiographische Skizze," in *Der gerechtfertigte Haeckel,* ed. Gerhard Heberer (Stuttgart: Gustav Fischer, 1968), p. 3–14. Reference is to pp. 8–9.

7. Uschmann, *Geschichte der Zoologie in Jena,* p. 41 n. 194.

8. Ibid.

9. Haeckel, "Autobiographische Skizze," p. 9.

10. Ibid.

11. Letter from Haeckel to Huxley, September 21, 1868, in Georg Uschmann and Ilse Jahn, "Der Briefwechsel zwischen Thomas Henry Huxley und Ernst Haeckel," *Wiss. Zeitschr. d. Friedrich-Schiller-Universität Jena,* Jahrg. 9 (1959–1960), Math.-Naturwiss. Reihe, Heft 1–2, pp. 7–33. The letter is on pp. 16–17.

12. Haeckel's "Author's Preface to the English Edition," *The History of Creation*, anonymous translation revised by E. Ray Lankester, 2 vols. (New York: Appleton, 1884), 1: xiii.

13. Uschmann, *Geschichte der Zoologie in Jena*, p. 45.

14. Charles Darwin, *The Descent of Man, and Selection in Relation to Sex*, 2d ed. (London: John Murray, 1877), p. 3 (emphasis added).

15. Uschmann, *Geschichte der Zoologie in Jena*, pp. 198–200, and plate facing p. 200.

16. J. H. F. Kohlbrugge, "Das biogenetische Grundgesetz. Eine historische Studie," *Zool. Anz.* 38 (1911): 447–453.

17. Charles Darwin, *On the Origin of Species* (London: John Murray, 1859), pp. 129–130.

18. Darwin's original draft for the chapter on natural selection has recently been published as part of R. C. Stauffer, *Charles Darwin's Natural Selection: Being the Second Part of His Big Species Book Written from 1856 to 1858* (Cambridge: Cambridge University Press, 1975). This version of the chapter is accompanied by several diagrams in which the upper-case letters representing the original species are located at the top of the diagrams, and the lower-case letters that represent those that diverge from them descend toward the bottom of the diagrams. Thus they bear no remote physical resemblance to an image of a tree.

19. Haeckel, *Natürliche Schöpfungsgeschichte*, 3d ed., p. 604.

20. Ibid., pp. 45, 598.

21. Haeckel, *Anthropogenie*, pp. 356–362.

22. Ibid., p. 340.

23. W. His, *Unsere Körperform, und das physiologische Problem ihrer Entstehung* (Leipzig: F. C. W. Vogel, 1874), pp. 168–171. Some but by no means all of the doubts shed on Haeckel's veracity have recently been countered by L. Szyfman, "Les prétendues falsifications d'Ernst Haeckel et son triomphe sur ses défamateurs," *Bulletin biologique de la France et de Belgique* 113 (1980): 375–406. Szyfman does not even refer to His's criticisms.

24. Haeckel, *Anthropogenie*, Fig. F1, Plate IV, following p. 256.

25. Ibid., Fig. 83, p. 273.

PART TWO

Methodology

7

Cladistic and Paleobotanical Approaches to Plant Phylogeny

PETER R. CRANE
and
CHRISTOPHER R. HILL

The study of fossil plants has recently occupied a curiously estranged position in relation to clarifying plant phylogeny. Although fossils provide the "genuine historical data" required by many who are interested in phylogeny (Sporne 1959; Davis and Heywood 1963), paleobotanical work has only rarely been directed toward specific phylogenetic objectives. Paleobotanists have mostly presented phylogenetic discussions in relatively broad terms or have been reluctant to pursue the phylogenetic implications of their data. One widespread view is succinctly expressed by Harris et al. (1974, p. 85): "It was said many years ago that the phylogenetic tree had become less like a tree than a bundle of sticks. It is easy by using the imagination to build the sticks into a coherent tree but the exercise is no longer held in esteem, for we know it can be done in many different ways." The conversion of "sticks" into a "tree" requires a perception of "relationship," and this is the fundamental issue in all systematics. At the operational level it influences both the techniques and the products of classification, and in evolutionary biology the notion of phylogenetic relationship is central to the recognition of evolutionary pattern and contributes to understanding the processes through which life evolves.

Cladistics adopts a concept of relationship that may be directly and meaningfully interpreted in evolutionary terms and that is the basis for its utility in phylogenetic and paleobotanical studies (Hill and Crane 1982). This chapter outlines the principles and methods of cladistics and discusses their implications for paleobotanical research with phylogenetic objectives. A cladistic classification of selected early land plants is used to illustrate two simple kinds of predictions that can be made from cladograms and that may be examined using the fossil record.

Principles and Methods of Cladistics

Cladistics is a method of hierarchical biological classification based on the observation that characters distributed among organisms can be perceived as exhibiting a hierarchical pattern. This pattern can be viewed as defining increasingly restricted, internested groups of organisms. For instance, seed plants (spermatophytes) form a restricted subclass of plants with tracheids (tracheophytes), which in turn is a restricted subclass of green plants. The hierarchical arrangement of characters and the groups within groups that they define leads to the expression of relationships in relative terms (Hennig 1965, p. 98). *Cycas* and *Ginkgo* both have seeds and are more closely related to each other than either is to the fern *Dryopteris,* which is free-sporing and lacks seeds. In practice, the pattern of character distributions among organisms rarely reveals a single simple hierarchy, but rather several conflicting patterns, depending on the characters chosen. Since reference is generally made to as many characters as possible, the principle of parsimony is used to choose between conflicting patterns. Parsimony minimizes the number of character conflicts, effectively minimizing the number of subclasses that overlap (Wiley 1981, p. 111). All the characters of the taxa under consideration may be useful for defining groups at some level in the hierarchy (Wiley 1981, p. 126), but at any given level of universality some characters will have definitional value in cladistics while others will be relevant at more general or more restricted levels in the hierarchy. Thus the possession of tracheids is of no significance in determining cladistic relationships within seed plants, but at the more general level of examining the relations of major plant groups it is of considerable importance and defines the more inclusive group Tracheophyta. Conversely, while possession of ovules enclosed at anthesis may be considered to define flowering plants, the occurrence of ovules without such enclosure in other seed plants (formerly grouped as gymnosperms) is of no cladistic significance. This concept of universality reduces to determining the utility of characters for group definition at any given level of analysis and is the fundamental difference between phenetics and cladistics. In effect, at all levels of universality in the hierarchy the phenetic concept of "overall similarity" is separated in cladistics into generalized and less generalized components, which have been termed symplesiomorphy and synapomorphy (Hennig 1965, 1966). These are simply the characters that are, respectively, irrelevant or relevant to the cladistic group-forming process at a given level in the hierarchy. Operationally, these hypotheses of relative universality are ultimately supported or rejected solely in relation to parsimony (Farris 1982). The problem of determining character polarity is nothing more or less than assessing the generality of character distributions for the purpose of determining which characters are nested within which. In practice, two approaches—"outgroup comparison" and "ontogenetic criteria,"—have been widely used to generate initial polarity hypotheses.

Outgroup comparison operates by raising the level of universality and placing the problem at hand in a broader, but still restricted, context (Stevens 1980; Watrous and Wheeler 1981; Hill and Crane 1982, p. 280). Thus, to determine whether proximal or distal spore germination is generalized (symplesiomorphous) in seed plants, the level of universality is raised to include other spore-bearing plants as outgroups. These exhibit proximal germination and support the hypothesis that this is a generalized feature. The ontogenetic criterion operates by extending the analysis to include the additional diversity of early ontogenetic stages (Wheeler 1981, p. 303; Watrous 1982, p. 99; but see also Nelson 1978, p. 327, for a different view). Thus, to determine whether needle-like or scale-like leaves are generalized in conifers, the analysis is broadened to include examination of seedling leaves. In many scale-leaved conifers these are needle-like, supporting the hypothesis that needle-like leaves are the generalized condition. Similarly, the narrow stomatal guard cells in the Gramineae pass through a phase in their ontogeny where their morphology resembles that of the more typical "kidney-shaped" guard cells of higher plants (Flint and Moreland 1946). This suggests that possession of reniform guard cells is the generalized condition. From hypotheses generated by these two methods, the application of parsimony minimizes conflicts between hypothesized synapomorphies that define groups (Wiley 1981, p. 111). In strictly operational cladistic terms the "true synapomorphies" that emerge are "homologies," and the problems of determining polarity, synapomorphy, and homology can be viewed in these terms as synonymous (Eldredge and Cracraft 1980, pp. 35–40; Hill and Crane 1982, pp. 282–287; Patterson 1982b).

The product, then, of cladistic analysis is a hierarchical scheme of nested synapomorphies in which the synapomorphies define classes or taxa. Like any other hierarchical scheme, it may be summarized graphically as a tree-like diagram, though there is no logically necessary tree-like form, temporal dimension, or relation to evolutionary theory (Nelson 1979, p. 8; Platnick

1979, 1982; Patterson 1982a). Such diagrams can in principle be fully independent of any phylogenetic preconceptions, even that evolution has occurred at all. They can be viewed merely as hypotheses concerning a perceived hierarchy of characters of organisms (Cracraft 1979, p. 30). Most systematists, however, wish to interpret and use such schemes in evolutionary terms, and historically the motivation for Hennig's development of cladistic techniques was the means they provided for investigating phylogeny (Hennig 1965, 1966).

Viewed in evolutionary terms, a sequence of synapomorphies of any set of internested taxa may be interpreted as a sequence of evolutionary novelties acquired successively in time from the bottom of the cladogram to the top. Relationships may thus be viewed in terms of relative recency of common ancestry (Bigelow 1956; Hennig 1966). The seed plants *Cycas* and *Ginkgo* share a more recent common ancestor than either does with the fern *Dryopteris*. The principle of parsimony can be viewed as a criterion for selecting from several competing cladograms that in which homoplasy (parallelism, convergence, and reversal) is minimized. The nodes of the cladogram can be interpreted as hypothetical ancestors, their characters specified by the character distributions defining the relationships of the terminal taxa (Hill and Crane 1982).

Traditional Paleobotanical Approaches to Phylogeny

Paleontological data, particularly in zoology, have traditionally been considered to have a fundamental role in elucidating phylogenetic history. In botany such information has figured much less conspicuously, and a recurrent theme has been that the absence or lack of sufficient fossil evidence poses an insuperable barrier to understanding plant phylogeny (Turrill 1942; Lam 1959; Davis and Heywood 1963). The central problem of exactly how fossils might reasonably be expected to contribute has received surprisingly little attention. One implicit assumption has been that phylogenies may be simply "dug from the rocks," with negligible theoretical input. The reasoning seems to have been that if enough rocks are split and enough fossils collected, the history of life will self-evidently be revealed (Schaeffer et al. 1972, p. 38). Such a view misses the crucial point that, even with a relatively complete fossil record, fossils can make phylogenetic sense only if they can be incorporated as part of the systematic and phylogenetic framework that other organisms, both living and fossil, provide. A perception of "relationship" cannot be avoided (Hill and Crane 1982).

Perhaps the notion contributing more than any other to the view that paleontology is central to comprehending phylogeny is that phylogenetic relationships are best expressed in ancestor-descendant terms. Phylogenetic cladistics offers an alternative in which relationships are expressed in terms of relative recency of common ancestry. In view of the low probability of recognizing individual ancestral species in a demonstrably incomplete fossil record, such an approach seems realistic. The response of traditional paleontologists to this problem has been to recognize "ancestral" higher-level taxa rather than species, on the basis that if it were possible to identify a precise ancestral species, then that species would fall within the higher group. Cladistically, this approach creates difficulties for defining such "ancestral" higher-level taxa. They can be defined only phenetically and not by features unique to them, because by definition they must include some members of the group to which they gave rise. In cladistic terms such taxa are paraphyletic and are effectively meaningless groups. For example, if the flowering plants evolved from a species included in the "gymnosperms," then the flowering plant + "gymnosperm" group (seed plants) comes to have one set of defining features, while flowering plants form a less inclusive subclass defined by other characters. The "gymnosperms" thus have no defining characters of their own. They can be diagnosed only in terms of a combination of generalized and less-generalized features. Other "ancestral" higher-level taxa, such as the "pteridosperms," are even more nebulous and difficult to define.

In addition to detection of ancestor-descendant relationships, it has also been advocated that the stratigraphic evidence derived from paleontology should play a central role in determining the polarity of character change, and thus in cladistic terms the level at which that particular feature should be applied to group definition within the classification hierarchy (Crisci and Stuessy 1980, p. 130; Harper 1976, p. 185; Szalay 1977, p. 117). The main difficulty with this approach is that the stratigraphic age of a fossil is independent of any notion of relationship. The logical link between stratigraphy and hypotheses of relationship is provided only by characters. The characters of fossils are no different from those of living organisms, and the manner in which they can contribute to hypotheses of relationship is identical. However, it is of no phylogenetic value to know that dichotomously veined leaves predate reticulate veined leaves in the fossil record without also knowing something about the plants on which the leaves were borne, how they are related, and what their relationships are to other living and fossil plants (Hill and Crane 1982, p. 274).

Cladistics and Paleobotany

Interpretation of cladistic hierarchies in evolutionary terms offers possibilities for clarifying the role of paleobotanical data in elucidating plant phylogeny. An important property of the synapomorphy scheme generated by cladistic analysis of character distributions is that it may be constructed independently of stratigraphic or other paleontological data. For instance, initial exclusion of stratigraphic data from cladistic hypotheses permits it to be retrospectively applied as an independent source of corroboration or conflict. In effect, this is an explicit examination of Louis Agassiz's "threefold parallelism" between comparative morphology, ontogeny, and stratigraphy (Gould 1976; Patterson 1981). Where conflict emerges, both the cladogram and stratigraphic evidence are open to reinvestigation, since both rest on complex subsidiary hypotheses that afford ample opportunity for error (Hill and Crane 1982). With reference to evolutionary theory, the cladogram can be used to make several predictions concerning the phylogenetic history of the terminal taxa under consideration. These may then be tested by reference to current knowledge of the stratigraphic record. These predictions concern both the sequence of appearance of characters through geologic time and the combinations of characters likely to be encountered as further organisms, both living and fossil, are discovered.

The relative sequence of first appearance of characters in the fossil record is predicted by the relative levels of universality of synapomorphies in a phylogenetically interpreted cladogram. The predictions may concern the relative sequence of first appearance of the states of a single character (single morphocline), in which stratigraphy can be used merely to examine the extent to which primitive characters are also ancient. However, they may also concern the nested relations of different characters (different morphoclines). A cladogram of ten early fossil land plants that have figured prominently in discussions of early tracheophyte phylogeny (Figure 7.1) gives a result similar though not identical to that of Banks (1975) and others. The cladogram defines five synapomorphy sequences (Figures 7.1 and 7.2) that show complete congruence with the known order of first stratigraphic appearance of the characters in the ten taxa under consideration (Figure 7.3).

Cladograms also make specific predictions concerning the character combinations likely to be encountered as new taxa relevant to a given monophyletic group are discovered. For example, from the cladogram presented in Figure 7.1, one might predict the occurrence of fossil plants that exhibit the apomorphic conditions of characters 1, 2, and 3; the plesiomorphic conditions of characters 5, 7–9, 11–16; and one or two of the apomorphic conditions of characters 4, 6, and 10. Such a plant would be a relatively primitive member of the *Leclercqia, Asteroxylon, Zosterophyllum*, and *Sawdonia* clade. *Renalia hueberi* (Gensel 1976) conforms to these specifications. It is known to possess tracheids (character 1), spores with a triradiate mark (character 2), and indications of stomata (character 3), although unequivocal guard cells have

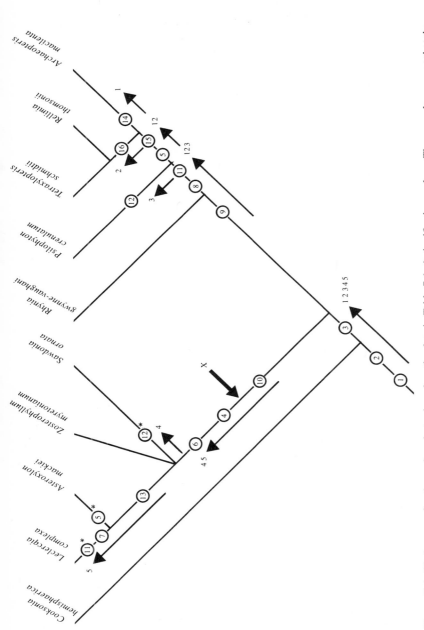

Figure 7.1. Cladogram of early tracheophytes based on the data in Table 7.1. * signifies homoplasy. These results suggest that the spinose enations (character 12) in *Sawdonia* and *Psilophyton* are not homologous, and that the regular, transverse sporangial dehiscence (character 10) in *Asteroxylon, Zosterophyllum* and *Sawdonia* is modified to longitudinal dehiscence (character 11) in *Leclercqia*. The cladogram suggests that longitudinal dehiscence in *Leclercqia* is not homologous with the longitudinal dehiscence found in *Psilophyton, Tetraxylopteris, Rellimia* and *Archaeopteris*. X indicates the position at which *Renalia* would be incorporated. Arrows and small numbers indicate the five principal character sequences.

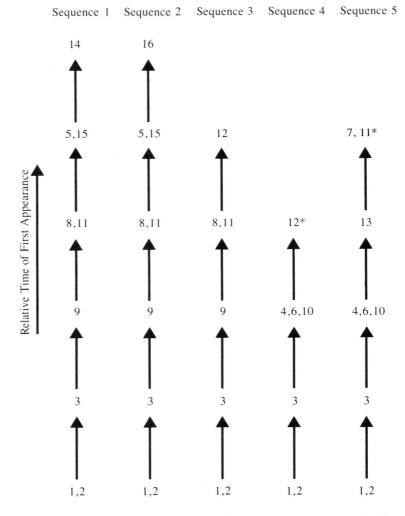

Figure 7.2. Five principal character sequences defined by the cladogram in Figure 7.1.

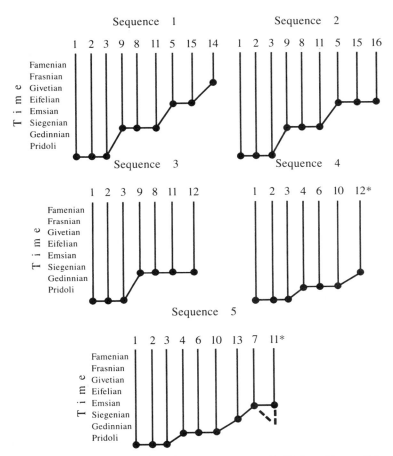

Figure 7.3. Comparison of the five character sequences (A-E) from Figures 7.1 and 7.2, with the first stratigraphic appearance of each character. Each sequence is arranged with characters ranging from more general to less general from left to right. Stratigraphic stage or series names are given on the left, and the first stratigraphic appearance of each apomorphic character in the ten genera under consideration is plotted. Stratigraphy follows Chaloner and Sheerin (1979), Banks (1980), and authors cited in the legend to Table 7.1. Vertical lines do not indicate stratigraphic range. The only apparent conflict between first stratigraphic appearance and the relative character sequence predicted from the cladogram is character 11* in sequence 5 (dotted line). The cladogram suggests that 11* in *Leclercqia* is not homologous to character 11 in *Psilophyton, Tetraxylopteris, Rellimia* and *Archaeopteris*. If *Leclercqia* is taken as indicating the first appearance of 11* the conflict with stratigraphy is eliminated (solid line).

Table 7.1 Data matrix for sixteen characters of ten fossil plants that have figured prominently in discussions of early tracheophyte phylogeny. With the exception of characters 7 and 14, unique features (autapomorphies) of the terminal taxa are omitted. Characters given in parentheses with question marks are inferred and have not yet been demonstrated. + indicates synapomorphies. N/A indicates characters not applicable. For characters 2, 4, and 15 in *Zosterophyllum myretonianum*, and 10 and 11 in *Rellimia thomsonii*, the conditions occurring in other species of the genus have been assumed. For character 3 in *Cooksonia*, stomata are assumed to be absent. It is also unknown whether *Cooksonia* possessed a cuticle. Following Chaloner and Sheerin (1979, text fig. 7), the xylem (characters 4, 5, 15) of *Cooksonia* has been assumed to be as in *Rhynia*. Microphylls are defined as regular, helically arranged leaves, with a single associated leaf trace that does not leave a leaf gap in the stele. In *Asteroxylon* this leaf trace terminates at the base of the microphyll. Note that tracheids in *Cooksonia* have been demonstrated only in sterile material (Lang 1937, p. 256; Edwards 1979, p. 32). Initial polarity decisions for characters 1–3, 8, and 10–16 were based on an increase in structural complexity as compared to living or fossil algal outgroups. Polarity decisions for the remaining characters (4–7, 9) have followed traditional comparative views. None of the polarity decisions is based on order of stratigraphic appearance, and in at least one case (character 4) stratigraphy marginally favors the reverse of the decision adopted (Chaloner and Sheerin 1979, text fig. 2). The final pattern is the simplest hierarchical explanation for the distribution of all the characters considered. Data from: *a*, Banks, Bonamo, and Grierson (1972); *b*, Kidston and Lang (1920); *c*, Lyon (1964); *d*, Hueber (1971); *e*, Gensel, Andrews, and Forbes (1975); *f*, Edwards (1975); *g*, Lang (1937); *h*, Edwards (1979); *i*, Kidston and Lang (1917); *j*, Edwards (1980); *k*, Doran (1980); *l*, Beck (1957); *m*, Bonamo and Banks (1967); *n*, Scheckler and Banks (1971); *o*, Leclercq and Bonamo (1971); *p*, Bonamo (1977); *q*, Phillips, Andrews, and Gensel (1972); *r*, Beck (1981).

Character	*Leclercqia complexa* Banks et al. -*a*	*Asteroxylon mackiei* Kidston & Lang -*b,c*	*Sawdonia ornata* (Dawson) Hueber -*d,e*	*Zosterophyllum myretonianum* Penhallow -*f*	*Cooksonia hemisphaerica* Lang -*g,h*	*Rhynia gwynne-vaughani* Kidston & Lang -*i,j*	*Psilophyton crenulatum* Doran -*k*	*Tetraxylopteris schmidtii* Beck -*l,m,n*	*Rellimia thomsonii* Leclercq & Bonamo -*o,p*	*Archaeopteris macilenta* Lesquereux -*q,r*
1. Tracheids absent, −; tracheids present, +	+	+	+	+	+	+	+	+	+	+
2. Spore wall lacking sporopollenin and a triradiate mark, −; spore wall with sporopollenin and a triradiate mark, +	+	+	+	+	+	+	(+)?	+	+	+
3. Stomata absent, −; stomata present, +	(+)?	(+)?	(+)?	+	−	(−)?	+	+	+	+
4. Endarch (centrarch) primary xylem, −; exarch primary xylem, +	−	−	−	−	−	(−)?	(+)?	+	+	+

Character									
5. Endarch (centrarch) primary xylem, –; mesarch primary xylem, +	N/A	+	+	N/A	–	N/A	+	+	+
6. Sporangia terminal, –; sporangia on short lateral stalks, +	+	+	+	+	+	–	–	–	–
7. Sporangia on short lateral stalks, –; sporangia on short lateral stalks borne on the upper surface of a microphyll, +	+	–	–	–	–	N/A	N/A	N/A	N/A
8. Terminal sporangia solitary, –; terminal sporangia clustered on profusely branched shoot systems, +	N/A	N/A	N/A	–	+	+	+	+	+
9. Sporangia spherical or broader than long, –; sporangia elongated, +	–	–	–	–	+	+	+	+	+
10. Sporangial dehiscence irregular, unspecialized, –; dehiscence regular, transverse, +	N/A	+	+	+	(–)?	N/A	N/A	(N/A)?	N/A
11. Sporangial dehiscence irregular, unspecialized, –; dehiscence regular, longitudinal, +	+	N/A	N/A	–	+	+	+	(+)?	+
12. Spinose non-vascularized enations absent, –; spinose non-vascularized enations present, +	–	+	+	+	–	+	–	–	–
13. Microphylls absent, –; microphylls present, +	+	–	–	–	–	–	–	–	–
14. Homosporous, –; heterosporous, +	–	–	–	–	–	–	–	–	+
15. Secondary xylem absent, –; secondary xylem present, comprising tracheids with circular bordered pits, and little parenchyma, +	–	–	(–)?	(–)?	–	+	+	+	+
16. Spores lacking a pseudosaccus, –; spores with a pseudosaccus, +	–	–	–	–	–	+	+	+	–

not been observed. The sporangia are broader than long with regular transverse dehiscence (character 10). For all other characters incorporated in Figure 7.1, *Renalia* does not exhibit (characters 6–9, 11–14, 16) or is thought not to exhibit (characters 4, 5, 15) the apomorphic condition. The characters of *Renalia* are therefore consistent with the predictions of the cladogram. Incorporation of *Renalia* as a sister group to *Leclercqia, Asteroxylon, Zosterophyllum,* and *Sawdonia* (X in Figure 7.1) dictates a definite sequence in which the characters of the lycopods *sensu lato* were acquired and leads to a more specific hypothesis of early tracheophyte evolution. In contrast to the situation in *Renalia,* the cladogram also indicates certain character combinations that would be unexpected, for example, terminal sporangia with regular transverse dehiscence, clustered on profusely branched shoot systems (apomorphic conditions of characters 8 and 10).

In effect, the cladogram predicts the limits and extent of mosaic character combinations, which may be tested against the mosaic of generalized and derived characters (heterobathmy) observed in fossil taxa. Corroboration of these predictions is significant, but fossil taxa are also of interest where they conflict with the expected patterns. In evolutionary terms, the cladogram attempts to provide a general but accurate hypothesis of phylogeny in a particular monophyletic group. If these requirements are satisfied, the cladogram should change very little, and certainly not fundamentally as new taxa are added. However, the addition of relevant fossils with character combinations contrary to those expected from the existing cladogram, as with the addition of any new taxon, may overthrow previous hypotheses of phylogenetic relationship. Clearly, such situations will be most common in monophyletic groups where much of the diversity is extinct.

Conclusions

The concept of relationship embodied in cladistics is based on mapping a character hierarchy perceived in nature (and by defining groups according to characters that they uniquely possess). Cladistic relationships may be viewed as phylogenetic relationships when that hierarchy is interpreted as resulting from evolutionary processes. Fossils have no special role in determining relationships, either through stratigraphy or through attempts to reconstruct ancestor-descendant lineages. It is equally clear, however, that adopting a cladistic approach is not antithetic to paleobotanical research; indeed, many phylogenetic discussions already existing in paleobotany can be directly interpreted in cladistic terms (Banks 1975; Doyle and Hickey 1976; Stein 1982). Criticisms of paleontology by some cladists have primarily been directed not to the record itself but more to the manner in which it has been employed for phylogenetic purposes. Fossils, like living organisms, are sources of characters and thus form part of the diversity that must be accounted for by any general hypothesis of phylogenetic relationships. The addition of further living and fossil taxa to an existing cladogram has the potential to alter the relative parsimony of different hypotheses of relationship. Cladistic methods currently provide the most explicit and highly resolved assessment of relationship which has a straightforward phylogenetic interpretation. They provide a highly useful framework in which to realistically address phylogenetic questions with paleobotanical data, and offer a theoretical context in which two of the major objectives of classical paleobotany—accurate documentation of stratigraphic appearance and the attempt to understand fossil plants as whole organisms—can make explicit and significant phylogenetic contributions. Phylogenetic hypotheses that are explanatory of both the available neontological and paleobotanical evidence (Miyazaki and Mickevich 1982; Crane 1985; Hill and Camus 1986) will be a prerequisite to examining some of the broader issues of contemporary evolutionary biology.

Acknowledgements

Discussions with Dr. Dianne Edwards, Dr. Patricia G. Gensel, and Dr. Andrew C. Scott concerning early tracheophytes are gratefully acknowledged. Dr. William Burger provided helpful comments on an earlier draft of this manuscript.

References

Banks, H. P. 1975. "Reclassification of the Psilophyta." *Taxon* 24: 401–413.
———. 1980. "Floral Assemblages in the Siluro-Devonian." In *Biostratigraphy of Fossil Plants,* ed. D. L. Dilcher and T. N. Taylor, pp. 1–24. Stroudsberg: Dowden, Hutchinson & Ross.
Banks, H. P.; Bonamo, P. M.; and Grierson, J. D. 1972. "*Leclercqia complexa* gen. et sp. nov., a New Lycopod from the Late Middle Devonian of Eastern New York." *Rev. Palaeobot. Palynol.* 14: 19–40.
Beck, C. B. 1957. "*Tetraxylopteris schmidtii* gen. et sp. nov., a Probable Pteridosperm Precursor from the Devonian of New York." *Amer. J. Bot.* 44: 350–367.
———. 1981. "*Archaeopteris* and Its Role in Vascular Plant Evolution." In *Paleobotany, Paleoecology, and Evolution,* ed. K. J. Niklas, pp. 193–230. New York: Praeger Press.
Bigelow, R. S. 1956. "Monophyletic Classification and Evolution." *Syst. Zool.* 5: 145–146.

Bonamo, P. M. 1977. "*Rellimia thomsonii* (Progymnospermopsida) from the Middle Devonian of New York State." *Amer. J. Bot.* 64: 1272–1285.

Bonamo, P. M., and Banks, H. P. 1967. "*Tetraxylopteris schmidtii:* Its Fertile Parts and Its Relationships with the Aneurophytales." *Amer. J. Bot.* 54: 755–768.

Chaloner, W. G., and Sheerin, A. 1979. "Devonian Macrofloras." *Spec. Pap. Palaeont.* 23: 145–161.

Cracraft, J. 1979. "Phylogenetic Analysis, Evolutionary Models, and Paleontology." In *Phylogenetic Analysis and Paleontology,* ed. J. Cracraft and N. Eldredge, pp. 7–39. New York: Columbia University Press.

Crane, P. R. 1985. "Phylogenetic Relationships of Seed Plants and the Origin of Angiosperms." *Ann. Mo. Bot. Gdn.* 72: 716–793.

Crisci, J. V., and Stuessy, T. F. 1980. "Determining Primitive Character States for Phylogenetic Reconstruction." *Syst. Bot.* 5: 112–125.

Davis, P. H., and Heywood, V. H. 1963. *Principles of Angiosperm Taxonomy.* New York: Van Nostrand.

Doran, J. B. 1980. "A New Species of *Psilophyton* from the Lower Devonian of Northern New Brunswick, Canada." *Canad. J. Bot.* 58: 2241–2262.

Doyle, J. A., and Hickey, L. J. 1976. "Pollen and Leaves from the Mid-Cretaceous Potomac Group and Their Bearing on Early Angiosperm Evolution." In *Origin and Early Evolution of Angiosperms,* ed. C. B. Beck, pp. 139–206. New York: Columbia University Press.

Edwards, D. 1975. "Some Observations on the Fertile Parts of *Zosterophyllum myretonianum* Penhallow from the Lower Old Red Sandstone of Scotland." *Trans. Roy. Soc. Edinburgh* 69: 251–265.

———. 1979. "A Late Silurian Flora from the Lower Old Red Sandstone of Southwest Dyfed." *Palaeontology* 22: 23–52.

Edwards, D. S. 1980. "Evidence for the Sporophytic Status of the Lower Devonian Plant *Rhynia gwynne-vaughani* Kidston and Lang." *Rev. Palaeobot. Palynol.* 29: 177–188.

Eldredge, N., and Cracraft, J. 1980. *Phylogenetic Patterns and the Evolutionary Process.* New York: Columbia University Press.

Farris, J. S. 1982. "Outgroups and Parsimony." *Syst. Zool.* 31: 328–334.

Flint, L. H., and Moreland, C. F. 1946. "A Study of the Stomata in Sugar Cane." *Amer. J. Bot.* 33: 80–82.

Gensel, P. G. 1976. "*Renalia hueberi,* a New Plant from the Lower Devonian of Gaspé." *Rev. Palaeobot. Palynol.* 22: 19–37.

Gensel, P. G.; Andrews, H. N.; and Forbes, W. H. 1975. "A New Species of *Sawdonia* with Notes on the Origin of Microphylls and Lateral Sporangia." *Bot. Gaz.* (Crawfordsville) 136: 50–62.

Gould, S. J. 1976. *Ontogeny and Phylogeny.* Cambridge, MA: Harvard University Press.

Harper, C. W. 1976. "Phylogenetic Inference in Paleontology." *J. Paleontol.* 50: 180–193.

Harris, T. M.; Millington, W.; and Miller, J. 1974. *The Yorkshire Jurassic Flora, IV: Ginkgoales and Czekanowskiales.* London: British Museum (Natural History).

Hennig, W. 1965. "Phylogenetic Systematics." *Annual Rev. Entom.* 10: 97–116.

———. 1966. *Phylogenetic Systematics.* Urbana: University of Illinois Press.

Hill, C. R., and Camus, J. 1986. "Evolutionary Cladistics of Marattialean Ferns." *Bull. Br. Mus. (Nat. Hist.) Bot.* 14: 219–300.

Hill, C. R., and Crane, P. R. 1982. "Evolutionary Cladistics and the Origin of Angiosperms." In *Problems of Phylogenetic Reconstruction*, ed. K. A. Joysey and A. E. Friday, pp. 269–361. New York: Academic Press.

Hueber, F. M. 1971. "*Sawdonia ornata*: A New Name for *Psilophyton princeps* var *ornatum.*" *Taxon* 20: 641–642.

Kidston, R., and Lang, W. H. 1917. "On Old Red Sandstone Plants Showing Structure, from the Rhynie Chert Bed, Aberdeenshire I: *Rhynia gwynne vaughani.*" *Trans. Roy. Soc. Edinburgh* 51: 761–784.

————. 1920. "On Old Red Sandstone Plants Showing Structure from the Rhynie Chert Bed, Aberdeenshire, III: *Asteroxylon mackiei* Kidston and Lang." *Trans. Roy. Soc. Edinburgh* 52: 643–680.

Lam, H. J. 1959. "Taxonomy: General Principles and Angiosperms." In *Vistas in Botany*, ed. W. B. Turrill, pp. 3–75. New York: Pergamon.

Lang, W. H. 1937. "On the Plant Remains from the Downtonian of England and Wales." *Philos. Trans., Ser.* B, 227: 245–291.

Leclercq, S., and Bonamo, P. M. 1971. "A Study of the Fructification of *Milleria (Protopteridium) thomsonii* Lang from the Middle Devonian of Belgium." *Palaeontographica, Abt. B., Paläophytol.*136: 83–114.

Lyon, A. G. 1964. "The Probable Fertile Region of *Asteroxylon mackiei* K. and L." *Nature* 203: 1082–1083.

Miyazaki, J. M., and Mickevich, M. F. 1982. "Evolution of *Chesapecten* (Mollusca: Bivalvia, Miocene-Pliocene) and the Biogenetic Law." *Evol. Biol.* 15: 369–409.

Nelson, G. J. 1978. "Ontogeny, Phylogeny, Paleontology and the Biogenetic Law." *Syst. Zool.* 27: 324–345.

————. 1979. "Cladistic Analysis and Synthesis: Principles and Definitions, with a Historical Note on Adanson's *Familles des Plantes* (1763–1764)." *Syst. Zool.* 28: 1–21.

Patterson, C. 1981. "Significance of Fossils in Determining Evolutionary Relationships." *Annual Rev. Ecol. Syst.* 12: 195–223.

————. 1982a. "Classes and Cladists or Individuals and Evolution." *Syst. Zool.* 31: 284–286.

————. 1982b. "Morphological Characters and Homology." In *Problems of Phylogenetic Reconstruction*, ed. K. A. Joysey and A. E. Friday, pp. 21–74. New York: Academic Press.

Phillips, T. L.; Andrews, H. N.; and Gensel, P. G. 1972. "Two Heterosporous Species of *Archaeopteris* from the Upper Devonian of West Virginia." *Palaeontographica, Abt. B, Paläophytol.* 139: 47–71.

Platnick, N. I. 1979. "Philosophy and the Transformation of Cladistics." *Syst. Zool.* 28: 537–546.

————. 1982. "Defining Characters and Evolutionary Groups." *Syst. Zool.* 31: 282–284.

Schaeffer, B.; Hecht, M. K.; Eldredge, N. 1972. "Phylogeny and Paleontology." *Evol. Biol.* 6: 31–46.

Scheckler, S. E., and Banks, H. P. 1971. "Anatomy and Relationships of Some Devonian Progymnosperms from New York." *Amer. J. Bot.* 58: 737–751.

Sporne, K. R. 1959. "On the Phylogenetic Classification of Plants." *Amer. J. Bot.* 46: 385–394.

Stein, W. E. 1982. "The Devonian Plant *Reimannia* with a Discussion of the Class Progymnospermopsida." *Palaeontology* 25: 605–622.

Stevens, P. F. 1980. "Evolutionary Polarity of Character States." *Annual Rev. Ecol. Syst.* 11: 333–358.

Szalay, F. S. 1977. "Ancestors, Descendants, Sister Groups and Testing of Phylogenetic Hypotheses." *Syst. Zool.* 26: 12–18.

Turrill, W. B. 1942. "Taxonomy and Phylogeny." *Bot. Rev.* 8: 247–270, 473–532, 655–707.

Watrous, L. E. 1982. Review of *Phylogenetics: The Theory and Practice of Phylogenetic Systematics,* by E. O. Wiley. *Syst. Zool.* 31: 98–100.

Watrous, L. E., and Wheeler, Q. 1981. "The Out-group Comparison Method of Character Analysis." *Syst. Zool.* 30: 1–11.

Wheeler, Q. D. 1981. "The Ins and Outs of Character Analysis: A Response to Crisci and Stuessy." *Syst. Bot.* 6: 297–306.

Wiley, E. O. 1981. *Phylogenetics: The Theory and Practice of Phylogenetic Systematics.* New York: Wiley.

8

Pattern and Process: Phylogenetic Reconstruction in Botany

PETER F. STEVENS

In this chapter, the focus is on the reconstruction of historical relationships between organisms, rather than on the production of a classification, with particular emphasis on problems that are acute for botanists. A major problem in systematics, particularly botanical systematics, is that evolutionary taxonomists working at higher taxonomic levels use a methodology of group formation that dates back to the latter part of the eighteenth century and a style of explaining relationships that was appropriate for that period. Relationships can be interpreted as if they are reticulate. However, cladistic (nonreticulate) relationships are nominally also being discussed, and this has led to tensions that contributed to the development of a cladistic approach to taxonomy.

As a systematist, I am interested in the theory and practice of pattern detection. I am also interested in explaining the patterns that I have recognized in terms of both the interrelationships of organisms and the process or processes underlying these relationships. These two aspects of systematics interact in a complex way: the generation of patterns cannot be entirely separated from a level of process explanation even when the latter is not known since we hope we are recognizing the pattern produced by the evolutionary process. Cladistic methodology, in many ways the most appealing methodology to use in the reconstruction of historical relatedness in botany, was developed by a zoologist, Willi Hennig. Problems encountered by botanists using this methodology are the main subject of this chapter, and the emphasis will be on relating these problems to the nature of the organisms being studied: plants rather than animals.

A brief sketch of the historical developments that have brought about a reevaluation in the way that we attempt to reconstruct phylogeny will put current practices and attitudes in a more general context. Then we turn to an outline of certain principles and premises of phylogeny reconstruction. Next I shall survey five areas currently under discussion: (1) use of outgroups in assessing which character states are derived; (2) weighting of characters; (3) homoplasy; (4) problems posed by "unresolved" structure in a cladogram (a branching diagram implying historical connections), in particular those associated with the recognition of ancestral taxa and of hybridization events; and (5) aspects of the relation between development, especially as manifest in organisms with basically indeterminate (or open) development such as plants, and cladogram construction. Discussion of many points is little better than perfunctory, but the references given allow entry into the literature. That I appear to be sceptical in certain places is more than a didactic ploy. Major advances in how we look at the world are still needed if we are to reconstruct historical relationships between plants in a satisfactory fashion.

The production of a classification from the pattern recognized, another integral aspect of systematics, will only be mentioned. However, differing goals as to what a classification should represent separate systematists such as Ernst Mayr and W. H. Wagner, who want both position of branching and amount of divergence to influence ranking and grouping of taxa, from their more cladistically oriented colleagues who allow only relative position of branching to influence rank (see, e.g., Young 1983). Associated with this difference are differing criteria for inclusion and exclusion of taxa in another taxon of higher rank. However, many taxonomists of both persuasions accept the importance of accurately detecting the relative positions of branching points in cladograms (e.g., Bremer and Wanntorp 1978), and the same methodologies may be used in the production of cladograms in both schools, since they have the common goal of reconstructing phylogeny (Mayr 1982, p. 230).

Historical Background

The acceptance of ideas of evolution in the 1860s and onward did not change the way taxonomists produced groups and classifications, but it did change the ways similarities between these groups were explained (for extensive discussion, see Stevens 1984a, 1984b). The hierarchical taxonomic patterns produced by both botanists and zoologists in the first half of the nineteenth century were an important element in the development of Darwin's thought (e.g., Ospovat 1981), and indeed the eventual success of aspects of Darwinian thought can be seen as largely vindicating the then-current taxonomic methodologies. Despite the common hierarchy used by botanists and zoologists, however, there were important differences in the way the two visualized relationships between members of that hierarchy. These differences developed in the first half of the nineteenth century and were associated with different ways of constructing groups and of seeing the world. They can in part be traced to differences in the nature of the organisms being studied (differences that still affect current discussions in systematics). They led to largely unperceived tensions in evolutionary systematics in botany during this century and, acting in conjunction with other factors mentioned below, to disillusionment with phylogeny as a goal of taxonomy.

In the 1830s, botanists working at higher levels in the taxonomic hierarchy were constructing classifications using data largely from external morphology, supplemented by some evidence from anatomy and less from paleontology and development. Knowledge of characters remained in some sense superficial, and it appeared that the "same" character occurred in a number of other taxa of varying ranks. Many of these character similarities were interpreted as evidence of relationships that in turn were seen as being largely reticulate. This is despite the hierarchical nature of the classifications produced and in the apparent absence of any general belief that the taxa were of hybrid origin. Typology, in the sense of the use of abstract types, was on the whole poorly developed (see also Stevens 1984a), and relationships were considered to be directly between extant members of the organisms or groups being discussed, rather than indirectly, via the type. Group formation was by establishment of some sort of similarity between members supposed to be related. Most groups were barely even polythetic, lacking either necessary or sufficient characters that would indicate group membership (see Beckner 1959; Sokal and Sneath 1963).

Evolutionary ideas might be supposed to encourage notions that relationships of taxa in a hierarchical system were mediated by an extinct ancestor or type (depending on how one thought), but such notions did not become strongly developed in botany. Relationships were still discussed as if they were reticulate and directly between groups, and principles of group formation remained unclear (Stevens 1984b, 1986). Different authorities could and did readily justify strongly differing circumscriptions of groups.

In zoology, on the other hand, evidence from anatomy, embryology, and even paleontology was coming to be widely used by the middle of the nineteenth century, particularly in the groups of more complex animals. Both the circumscription and coherence of many of these groups became more apparent, and associated with this was the strong development of typological thinking, a new development in taxonomy then (Cannon 1978). Relationships between groups were more clearly hierarchical and nested and were sometimes depicted as being tree-like, and these relationships were discussed as being indirect, via the type. This method of visualizing relationships was largely compatible with evolutionary thought, despite heated argument as to whether a type was an ancestor or not (constitutive vs. regulative types; see

Lenoir 1978). However, even to some zoologists any evolutionary reconstructions were tentative and suspect, and especially toward the middle of the twentieth century, trees became more shrub-like (Jepsen 1944).

In both botany and zoology there were problems with the apparently logically circular relationships between morphology, classification, and evolutionary thought. Fossils would provide clear-cut evidence of evolutionary relationships, yet the fossil record was too fragmentary to be of much use. Additional problems arose with the subjective nature of decisions as to what were (evolutionarily) important characters and as to the ranking of taxa. Claims that taxonomy was not scientific were countered by Simpson's (1961) declaration that the taxonomic process was largely in the nature of an artistic process, and hence by implication was outside the bounds of scientific criticism. Another reaction was the development of a phenetic approach to group formation, in which an attempt was made to use "all" characters without selecting or weighting particular characters and divorcing the process of group formation from evolutionary considerations. By these means taxonomy was to become both objective and scientific (see Hull 1970, for a useful summary).

A third reaction was an attempt to construct phylogenetic classifications themselves on a more scientifically defensible basis. In zoology the name of Hennig is particularly associated with these developments; in botany it is the name of Wagner (see Hennig 1966; Wagner 1952). Both developed an explicit methodology for using shared derived characters alone (synapomorphies) in the recognition of branching points and lineages; at one level these can be considered evolutionary novelties arising in species that are ancestral to the lineages. The main difference between the two is in the principles of formal classification. Hennig (1966) considered that all taxa must be monophyletic, consisting only of all known descendants of a single stem species; the amount of change along a lineage did not affect taxon rank. Wagner, on the other hand, emphasized amount of change as affecting both the circumscription and the ranking of taxa and adopted a different definition of monophyly (similar to the definition of convexity adopted by other evolutionary systematists, and including the paraphyletic taxa of Hennig; see Wiley 1981b).

Outline of Cladistic Analysis

The first step in cladistic analysis, as in other methods of data analysis, is recognition of the basal units to be studied, an operation I shall not discuss here. Characters are then recognized by similarity of structure in different organisms, and they are subdivided (for the purposes of discussion a character can be any attribute of an organism). These are obviously critical steps. The first is mentioned often in the literature (for references, see Kaplan 1980;

Sattler 1984; Stevens 1984c); it should be emphasized that the criteria used to judge similarity are not unambiguous. Many of the arguments about relationships in botany hinge on whether a character in one group is "really" the same as a similar character in another group. Not surprisingly, different answers provide different ideas of relationships (for an example, see Cartmill 1982). The other step, subdivision of a character (e.g., fruit a 5-locular capsule vs. fruit a 3-locular capsule), is usually little analyzed, but incorrect or unjustifiable division of characters is probably widespread in botanical literature. Indeed, justification of this step is rarely given (but cf. Hennipman and Roos 1982 and Almeida and Bisby 1984, for the methodology to be followed). Occasionally an author simply expresses the hope that no problem will be encountered, at least if qualitative characters are used (LaDuke 1982). However, even qualitative differences may simply represent semantic discontinuities. The distinction between leaf bases cordate or rounded, or undersurface of the leaf glaucous or not, does not reflect necessary discontinuities in the character itself (see also Hart 1983); the terms are applied to part of a potential continuum of variation. Terms denoting colors reflect the same problem. Furthermore, in this case the human eye is more sensitive to certain colors, so semantic discontinuities are underlain by perceptual discontinuities, but none of this bears on the "reality" of the colors in nature. The net result of incorrect subdivision of characters is to vitiate or simply render circular any subsequent character analysis (Stevens 1984c). When one reads many lists of characters (e.g., Judd 1981; Bolick 1981; Sanders 1981; Whalen, Costich, and Heiser 1981; Funk 1982; Baum 1983), it often appears that the subdivisions recognized may indeed represent arbitrary divisions of a continuum. Any analysis of such characters will be problematic.

The characters recognized are then polarized, one subdivision being recognized as more general, primitive, or plesiomorphous than another or others that are less general, derived, or apomorphous. Such assignments are best made by looking at the characters in the sister group of the study group and ideally also the sister group of that combined group (see Stevens 1980a and Wheeler 1981 for details and possible alternative criteria). By this method of character analysis, if a character shows expressions a and b in the ingroup, but consistently only a in the outgroups, then a is primitive and b is derived. Such decisions follow from parsimony considerations (e.g., Stevens 1981; Farris 1982). Outgroup analysis, although formally simple, may present problems in practice and is discussed below.

Subsequent to this stage of analysis, a variety of manual or computer-assisted algorithms are applied to the data, and a branched diagram, or dendrogram, is produced. In such a diagram, characters are displayed across taxa in such a way that character changes are reduced to a minimum (consistent with the constraints of the algorithm). Thus, given a data set such as that in

Figure 8.1a, the distribution of characters in taxa can be depicted as in Figure
8.1b, in which there are a minimum of six changes of characters. An alternative explanation, Figure 8.1c, involves nine changes, with the relationships
between taxa poorly resolved (see also below). Care must be taken to use algorithms based on realistic assumptions about character changes. Such assumptions have long been discussed (e.g., Funk and Stuessy 1978), but it is
not always clear that they are treated as more than trivial matters. For example, algorithms that do not allow reversal of characters will never give patterns supporting process explanations that permit reversals.

	A	B	C	D
1	1	1	0	0
2	0	0	1	1
3	1	1	0	0
4	0	0	1	1
5	0	1	0	1
6	1	0	1	0

a

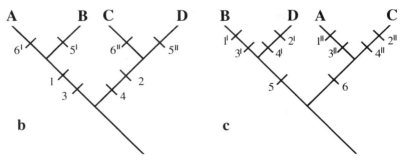

b   c

Figure 8.1. Parsimony and homoplasy. (a) Data matrix: A–D, taxa; 1–6, characters;
0, 1, derived and primitive states. (b) Cladogram displaying most parsimonious arrangement of data (initially to be preferred): characters 5 and 6 homoplasious, total
of eight character state changes (i.e., 0 → 1). (c) Cladogram showing alternative
arrangement of data: characters 1, 2, 3, and 4 homoplasious, total of ten character
state changes.

A distinction can be made between the production of a dendrogram, a cladogram inferring historical connections, and that of a phylogenetic tree in which biological relationships between taxa are specified (e.g., Wiley 1981, p. 97; see also Nelson and Platnick 1981, p. 171). There has been much discussion over whether the methodological parsimony involved in production of a dendrogram implies evolutionary parsimony in the context of discussions of evolutionary trees. The answer is that hypotheses about phylogeny, like other hypotheses in science, are basically the simplest that can be proposed based on the data at hand, and they explain all the data with the fewest possible extra assumptions (see Beatty and Fink 1980; Wiley 1981; Farris 1982, 1983; Brady 1983; Sober 1983). More complex hypotheses can be accepted only if there is evidence in their favor, when they in turn better reflect the data. Parsimony (or simplicity) is not an easy tool to use, and this becomes apparent in the discussion of homoplasy and weighting.

Problems of "Cladistic Analysis" in Botany

CHOICE OF OUTGROUPS

One of the problems with the use of outgroup analysis has been that the monophyly of the study group and its sister taxon must be established (Stevens 1980a and references). More careful recent reformulations of the outgroup rule have shown that ideally one should know the sister group of the two taxa combined as well (Maddison, Donoghue, and Maddison 1984). How can one know all this when what one is trying to do is work out, maybe for the first time, phylogenetic relationships? This problem has often been raised (e.g., Funk 1981; Sanders 1981; Young 1981; Baum 1983). An example from the Ericaceae (the heath family) will illustrate the kinds of approach that can be followed.

A more or less classical approach allows the Ericaceae to be delimited as a fairly coherent morphological unit (Stevens 1971). The Ericaceae are recognized by a series of floral and embryological features which, although not unique in the angiosperms, are severally uncommon and collectively do not occur together except here. These include anthers that invert during development, generally dehisce by pores, lack an endothecium, have pollen in tetrads, and so on. There is another complex of characters from various parts of the plant, including indumentum, flower, and seed, which is also common in the Ericaceae and somewhat less commonly found as a complex elsewhere, although individually the characters may be quite common. On the basis of an evaluation of such characters by what might be called a cladistic approach with a phenetic overlay, that is, emphasizing the importance of possession of the uncommon features for membership in the family, the Empetraceae, a separate family, will be included in the Ericaceae, and the Ericaceae subfamily Wittsteinoideae will probably be excluded. This leads to recognition of a much more nearly monophyletic unit.

The immediate relatives of this unit, as well as details of its internal cladistic structure, can be elucidated by using individually as outgroups the various groups that other authors have indicated as relatives—the outgroup substitution method—much as suggested by Cantino (1982) and Donoghue (1983b). However, as Cantino (1982) and Barabé (1982) have found, the larger the outgroup the more difficult it is to polarize characters because of homoplasy (see also below). The outgroup can be analyzed at the same time as the ingroup (Maddison et al. 1984), but care should be taken to use the same polarity criteria in the analysis of the outgroup as in analysis of the ingroup (cf. Kress and Stone 1983).

As structure within the Ericaceae appears, the groups delimited can then be used as functional outgroups in further analysis within the family (see Watrous and Wheeler 1981). This should not be interpreted as becoming locked into a potentially erroneous cladistic structure, since a continuing part of the taxonomic process is continual reevaluation. As a taxonomist comes across new taxa and characters in the course of work, he or she automatically and almost without thinking compares them with familiar taxa and characters. It was by this means that I first realized that *Alseuosmia*, of the Alseuosmiaceae, and *Wittsteinia*, of the Ericaceae, were possibly more related to one another than to any other taxa. Of course, monographic work should as a matter of principle involve a total reevaluation of a group (see Semple 1981). Hence the argument against outgroup analysis that it requires *knowledge* of phylogeny (Guédès 1983) is overstated, for "knowledge" in systematics is acquired in a complex and nonlinear fashion (Hull 1967).

HOMOPLASY

Figure 8.1b shows that character 5 changes twice. Characters that occur more than once in a cladogram are called homoplasies, or parallelisms. In many botanical analyses, homoplasies are very numerous, and sometimes all the characters used to indicate relationships are homoplasious (Riggins and Farris 1983; see also Funk 1981, 1982; Donoghue 1983a). Four points may be made in connection with this.

1. Homoplasies are interesting objects of study in their own right, but they can only be recognized with reference to a phylogenetic framework. Even if the occurrence of homoplasies is believed to result from a (presumably inherited) tendency to develop a certain character (e.g., Cantino 1982), such a

tendency can be suggested only if an entirely different set of characters is used to construct the cladistic framework. Otherwise the argument becomes dangerously circular (in much botanical evolutionary taxonomy, parallelisms are taken as evidence of common ancestry; Stevens 1986).

2. It is clearly a feeling among some, perhaps many, botanists that the high homoplasy levels are intrinsic to plants because of the nature of their growth and development (Sanders 1981; see below) and may make cladistic analysis difficult. However, such analysis does not become impossible when homoplasies outnumber nonhomoplasious characters (cf. Sanders 1981). Homoplasies will probably support several competing cladograms, since they are of independent origin, while nonhomoplasious characters reflecting phylogeny will all support a single tree (see also Funk 1982). Sober (1983) outlines evolutionary conditions under which cladistic analysis is possible; the rate of origin of a character is lower than the rate of its transmission from species to species (this can be modified to deal with hybridization). Felsenstein (1978b) has suggested that at least some of the algorithms used in cladistic analysis may show worrying properties when homoplasies are common and inequalities in rates of evolution within the study group become large.

3. Cladistic analysis in some plant groups results in cladograms with a structure that is very vulnerable to changes in the coding of even a single character, let alone addition or removal of taxa or suites of characters (Coombs, Donoghue, and McGinley 1981; Donoghue 1983a). A similar lability is suggested by the observation that dendrograms based on data from different organ systems show poor agreement (see also the nonspecificity hypothesis; Sneath and Sokal 1973). Another aspect of this vulnerability to new data is that in analyses of large groups, which are often also those with much homoplasy, a number of trees of approximately equal length (number of character changes) but different structure may be produced. Simple acceptance of *the* most parsimonious tree in such a situation would be equivalent to "bean bag genetics." Trees must be evaluated, just as much as characters may have to be weighted (see also below) to see which, if any, tree is most likely. Given our understanding of organisms and characters, it may prove impossible to select any one tree, but stable parts common to all trees (congruent, robust) can suggest relationships that may be worth discussing (see also Coombs et al. 1981; Donoghue 1983a). Or, perhaps less usefully, consensus trees, effectively the lowest common denominator of the patterns of the various trees, may be computed (Adams 1972). However, in the analysis of major groups, as in the attempt to disentangle vessel evolution in the "primitive angiosperms" (Young 1981), most parts of the tree may be poorly supported by synapomorphies, homoplasy may be rife, and the characters available may not provide reliable evidence of phylogenetic relationships (Riggins and Farris 1983).

4. One of the main endeavors of taxonomists generally should be to im-

prove the quality of their basic data. With more careful analysis and improved methods of studying plants, it is to be expected that differences will be found between structures initially thought to be similar. The problem of homoplasy may in part be a pseudoproblem. If it is not, massive amounts of homoplasy may indeed compromise attempts at cladistic reconstruction.

WEIGHTING OF CHARACTERS

The problem of weighting of characters was a major spur to the flurry of methodological and philosophical developments in systematics beginning in the late 1930s. It is a problem that has not been solved and has not gone away, and it affects all schools of taxonomy (Davis and Heywood 1963; Mayr 1969; Sneath and Sokal 1973; Kluge and Farris 1969).

Circularity problems are inherent in character weighting, as are problems in developing defensible criteria for weighting in one case and not in another. It is unwise to assume that some characters in plants *cannot* arise in parallel, or even reverse, despite the reliance that has traditionally been placed on certain suites of characters such as xylem anatomy, chromosome number, or floral characteristics in evolutionary studies (see Stevens 1980a for references). One might wish to evaluate equally parsimonious cladograms by rejecting some characters and seeing what the resultant structure will be when the subset of data is reanalyzed. The criteria used in rejecting characters from the data set may be prior experience with a character, including knowledge of its development and its distribution in a variety of groups, or dissatisfaction with the cladogram because of the distribution of characters that had not been used in the initial analysis (Coombs et al. 1980; Donoghue 1983b; Gosliner and Ghiselin 1984). However, the consequences of rejection of a character simply because it supports a cladogram that conflicts with one's phylogenetic preconceptions are all too evident.

Probabilistic notions of change in character state could be introduced but these would conflict with our desire to know about particular cases and could anyway be developed only from a knowledge of many particular cases. Weighting a character according to its apparent complexity, that is, breaking it down into a whole series of independent characters, is possible, but even apparently less complex characters may be amenable to such a breakdown, and our notions of complexity or evolutionary importance of a character are notoriously unsatisfactory. A similar form of weighting occurs when a character in a group appears to show a complex transformation series, for example, a_0—a_1—a_2—a_3—a_4. Assuming that a_0 occurs in the outgroup, the character could be treated so as to give the character a_4 a weight four times that of a_1: a_1 would be scored 1000; a_2, 1100; a_3, 1110; and a_4, 1111 (additive binary coding; see Figure 8.2d). More complex arrangements of the variation shown

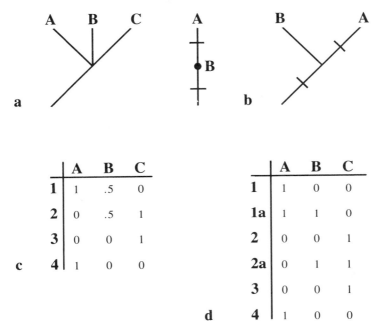

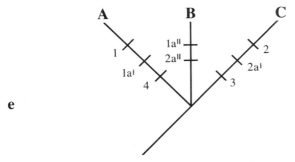

Figure 8.2. Ancestors and hybridization. (a) Trichotomy. (b) Alternative representations of ancestor-descendant relationships: bars represent apomorphies or suites of apomorphies. (c) Data matrix of hybrid and two parents: A–C, taxa; 1–4, characters, characters 1 and 2 showing characters intermediate in expression between parents, characters 3 and 4 representing characters expressed only when homozygous. (d) Data matrix converted into two state characters (additive binary coding): characters showing simple dominance will be scored like characters 1a and 2a. (e) Cladistic representation of data matrix in (d): characters 1a and 2a showing apparent homoplasy.

by characters can be imagined; in all cases the evolutionary progressions that it shows are being assumed. If all the character subdivisions were treated independently, this would again imply that the evolution of this character complex was known, with each subdivision representing an independent transformation of a_0. A solution to this form of weighting might be to treat all such complex characters as a single, binary character in the initial analysis: a_0 versus a_1—a_4.

UNRESOLVED STRUCTURE

As Hennigian methodology was initially formulated, there was a nearly explicit assumption that speciation consists of a splitting of an ancestral population (species), with both new populations developing apormorphies. The final, resolved structure of any cladogram resulting from such a process will then consist entirely of dichotomies and terminal taxa. Some authors since have suggested that such a structure will be attained when cladistic resolution is complete: nature has a hierarchical structure (e.g., Platnick 1979; Rosen 1982). Examination of two kinds of "unresolved" structures possible in a cladogram, a trichotomy, and an apparent ancestral taxon (Figure 8.2a and b) reveal conflicts in some approaches to phylogeny reconstruction using Hennigian methodology and also will lead into a discussion as to what cladistic pattern might be expected to represent in terms of representation of process. It will also help focus on the level at which Hennigian methodology is appropriate.

HYBRIDIZATION

The pattern in Figure 8.2a might result from a hybridization event. What are the conflicts between a methodology that initially largely ignored hybridization, and its use in the analysis of groups in which hybridization is suspected to occur? (Cladistic methodology was developed by a zoologist, and a number of zoologists to this day discount hybridization as being important in animals, while ignoring botanical problems.) The role of hybridization as a taxon-generating mechanism has been empirically established, and it is probably of widespread occurrence (cf. Brady 1983). Evidence for or speculations and references about the importance of hybridizations in ferns, gymnosperms, and flowering plants are found in Grant (1981), Ehrendorfer et al. (1968), Löve, Löve, and Pichi Sermolli (1977), Knobloch (1976), Wagner (1983), and many other publications. Strangely enough, critical reaction to the problem has been slow in developing among botanists (but see Bremer and Wanntorp 1979). Usually an attempt has been made simply to remove the hybrids from the analysis (e.g., Sanders 1981, Wagner 1980, Funk 1981,

Humphries 1979). Recent books on cladistic methodology either largely avoid the issue (Wiley 1981a) or state that "reticulate trees do not integrate information; they merely reproduce it" and are not appropriate as possible candidates for the best cladogram or generally most parsimonious tree (Nelson and Platnick 1981, p. 218). However, the situation has recently been rectified (see esp. Wanntorp 1983, Humphries 1983). The problem of hybridization in classification is extensively discussed by Wagner (1983 and references). Although the extent of introgression, the transfer of genes from one taxon to another, has been questioned (e.g., Heiser 1973), it and viroid-mediated gene transfer pose problems similar to those being discussed for hybridization.

Many first-generation hybrids produced by a cross between parents of two different species (F1 hybrids) tend to have characters with expressions intermediate between those of their two parents. However, even in such "ideal" circumstances, even if the parents are sister species, and even if we recognize all the intermediate and parental character states as distinct, the hybrid may still be represented as a trichotomy (Figure 8.2e), which in itself gives no evidence of origin. The problem is far more severe, however, and is summarized in the following six points.

1. Unfortunately, there is no rule in nature that only sister species can hybridize. One may even argue, especially if one is a proponent of the biological species concept, that such taxa would perhaps be least likely to hybridize. This pattern has indeed been found (Rosen 1979 [see summary in Wiley 1981a] for *Xiphophorus,* swordtail fish). Hybrids between taxa not immediately related are almost commonplace in botanical analyses (e.g., Humphries 1983) and are connected with the capricious development of isolating mechanisms in relation to the speciation process in plants (see Levin 1978 for a summary).

2. Even F1 hybrids, which are often more or less infertile, are not necessarily phenotypically intermediate, so the doubling of chromosome number and restoration of fertility in an F1 hybrid will not necessarily yield a taxon that is recognizably intermediate between its parents. Furthermore, hybrid taxa may result from back-crosses with one of the parental species, after selection within a hybrid population or after crossing between two taxa that are themselves hybrid, and so forth (e.g., Grant 1981). Such taxa may have character states approaching or even identical to those of only one of their parents. Selection may act on members of the taxon or its progeny, and the longer ago that hybridization occurred, the greater the probability that the character states of the hybrid deviate from any intermediate condition and that one or perhaps both of its immediate parents have become extinct. It is of no comfort that some authors believe that hybridization and polyploidy do not generally

produce evolutionarily very important lineages (see Funk 1981 for references), since this belief may be incorrect in the case of the pteridophytes and perhaps some angiosperms.

3. At least in some cases, one does not know which taxa are of hybrid origin. After all, one generally does not know the cladistic relationships within a group before one begins an analysis. Hence characters will be determined only after examination of the whole group. There is no knowing how a character in a hybrid that is "intermediate" between that in its two parents will be analyzed when the whole group is considered. In addition, any intermediacy of character expression in a hybrid may make it difficult to divide up the variation in the entire group into discrete characters (see above). Even if taxa are compared in threes, when the taxon intermediate between the two others is suspected of being of hybrid origin (the triplex hybrid detection method), these problems (which can be circumvented if the character analysis itself is restricted to these three taxa) and those in the preceding paragraph (less easy to circumvent) may lead to hybrids remaining undetected (cf. Wagner 1983). Cladistic analysis is similarly blind, and thus no particular position can be expected of hybrids in the cladogram. They will not necessarily form trichotomies (cf. Funk 1981), and they will not necessarily be intermediate in general position in the cladogram if their parents are evolutionarily far apart (e.g., Humphries 1983, fig. 6.37). Serious loss of structure of the cladogram may result if conventions to avoid the use of conflicting characters are followed (see Bremer and Wanntorp 1979, Humphries 1983, cf. Funk 1981).

4. Hybridization increases the effective homoplasy of the data set. Although one might hope that "homoplasy" that is the result of hybridization will occur in a variety of characters that are not all related to one adaptive syndrome, it seems a rather forlorn hope to distinguish between a set of homoplasies that result from hybridization and homoplasies that are the result of independent origin of a set of characters perhaps associated with a single adaptive syndrome. ("Intermediacy" may also result from differential fixation of alleles in descendant populations/species that were heterozygous in an ancestral population/species, a situation that is not the result of hybridization and is arguably not even due to homoplasy.)

5. It may be possible to produce more parsimonious trees if reticulation is allowed, as Nelson and Platnick (1981) have suggested. The underlying problem is still the pattern of characters: homoplasy and hybridization may both result in the same pattern of characters. In addition, with the allowance of reticulation events, the problem of computation of the most parsimonious cladograms, already difficult (Felsenstein 1978a), becomes much more complicated.

6. Hybridization confuses the issue of monophyly, because taxa involved in hybridization that results in species are no longer strictly monophyletic. However, this issue concerns classification rather than phylogeny reconstruction and is not discussed here (although it needs critical discussion; Hull [1979] is overly sanguine about the problem, which he basically considers in the context of hybridization of closely related species).

The fundamental problem is that one is trying to detect reticulation events while using a methodology that was initially developed under the effective methodological assumption that such events did not occur. Thus Brady (1983, p. 59) considers parsimony in the context of "the demand of hierarchical coordination and the particular data set." In addition, reticulation events are the result of a class, rather than a single kind, of events that have neither an invariant nor a unique pattern of character distribution associated with them and so will not be easily recognizable. As Wiley (1981a, p. 37) noted, "The nature of the evidence used to support hypotheses of such patterns [reticulation events] is rather different from that encountered in the usual analysis of clades forming simple (nonreticulate) phylogenetic trees."

Only general guidelines seem possible at this point. All taxa should initially be kept in the analysis. On examination of the cladograms produced, hybrid taxa may be suspected because they show similarities to two taxa/lineages, and thus they may change their position considerably depending on order of insertion of data, subdivision of data sets, and other manipulations of the data. Alternatively, their positions may not change, but a particular taxon may have many homoplasies with a taxon/taxa in other parts of the cladogram. Such "homoplasious" taxa can be integrated as reticulations into the cladogram by following the method suggested by Nelson (1983). They should be further examined, and an attempt made to establish their hybrid nature by examining new data, for instance, chromosome number, geographical distribution, and so forth, that were not examined initially, and by attempting to resynthesize the hybrid (e.g., Sanders 1981; Wagner 1983), that is, by using conventional methods of studying putative hybrids. Taxa still suspected of being of hybrid origin should then be removed from the analysis and a new cladogram computed. This new cladogram may show substantial differences from the first, especially if several hybrids are involved and/or the hybrids have substantially increased the homoplasy of the group.

Whether ancient hybrids will be detected is still very questionable. It does not seem reasonable to restrict attempts to reconstruct historical relationships to groups that do not show reticulation (cf. Kluge 1983). Rather, one should perhaps avoid the use of conventional cladistic methodology on such groups. As Humphries (1983, p. 103) suggested, hybridization and its recognition suggest that "we [seem] to have reached the limits of cladism"—at least any cladism that is unable to represent nonhierarchical relationships. Reticulate pattern produced by the hybridization process is part of the real world of those who work on phylogenetic reconstruction at many taxonomic levels in plants.

ANCESTRAL TAXA

In Hennig's initial methodological proposals, ancestral taxa also played a restricted role. A taxon in the fossil record could never change its name so long as its lineage did not divide, no matter how much its morphology changed. Once it divided, however, the name of the ancestral taxon could not be kept. Two new taxa would be produced, each with its own unique apomorphies and each presumably with its own unique name. In work on extant taxa, ancestral taxa have not been viewed kindly by some cladists, especially by those primarily interested in pattern (see Hull 1979 for references). The problem is when can a situation like that depicted in Figure 8.2b represent biological reality? Clearly, the initial methodological assumptions militate against its acceptance (Platnick 1979), but might it represent a biologically reasonable situation? If so, how can we recognize it? (For references to possible examples in *Clarkia, Stephanomeria,* and *Gaura,* see Jackson 1985.) Suppose a species consisted of a single population (totality of organisms exchanging genes), and then a new population was established from a single propagule of this species, and it developed its own apomorphies. In this case, there would be no necessary reason for the ancestral population to be characterized by *its own* apomorphies. This is a possibility that cannot be ignored. However, to confirm a pattern interpretable as that of ancestor-descendant relationship is equivalent to proof of a null hypothesis—an ancestral taxon has nothing unique to it, and its existence becomes likely only when evidence from a variety of sources fails to suggest other patterns. Furthermore, there is nothing to prevent the ancestral population from subsequently acquiring its own unique characters; two "sister taxa" then become recognizable. Although the pattern produced now becomes that of a vicariant event (splitting of a continuous range), the process that produced it has an important dispersal element. Again, pattern and process are only loosely connected in a cladogram.

It should be emphasized here that a large part of the ancestor-descendant problem may be a problem with the level of analysis. "Species" in many groups of plants may consist of multipopulational units of which the individual populations are indistinguishable at our current level of analysis. Such species are treated as units, and it is a commonplace in systematics to think of other species as being derived from such multipopulational species (e.g.,

Stevens 1980b, Whalen et al. 1981, Ashton 1982). Nothing distinguishes the individual parts of the larger species, but the species derived from one of those individual parts is itself distinguishable. Urgently needed are better data on the extent of effective gene flow (see Mishler and Donoghue 1982 for references) and closer analysis of conventional species.

Use of Developmental Data

There is currently much discussion, especially in zoological circles, about the role of developmental data in detecting hierarchical pattern. Here I discuss two aspects of this argument from a botanical point of view. The first concerns the nature of development in plants, the second concerns the possibility of detecting various kinds of developmental change.

Plant development is open, not closed. That is, many plants do not go through a sequence—young, adult, senescent—as do many animals, but basically reach an adult stage and then persist. The reproductive-adult/ nonreproductive adult stages follow each other, the reproductive parts being lost in their entirety during the nonreproductive stages. Sometimes there are even reversals to the young stage, or parts of the cycle may drop out. It is possible that such parts of the cycle may later be regained in all their former complexity (Basile and Basile 1983; see Stevens 1984c for additional references). Since most "higher" plants are attached to the substratum, extensive genotypically and phenotypically controlled modifications of the plant body may allow an individual to persist in a given locality despite changing environmental conditions that cannot be avoided.

The simplicity and apparent malleability of the plant body has long presented a problem to botanists. Thus we find Gray (1842), Hooker (1856), Engler (1892), and Bessey (1893), and others early bewailing the difficulty of distinguishing between what would now be called primitive simplicity and derived simplicity, the latter being simplicity produced by the loss of characters that the ancestor once had. Little in development suggests the existence of primitive vegetative complexity in parasitic organisms such as *Cuscuta* and *Cassytha,* which have an apparently simple vegetative body that looks like fine twine. Sattler (1979) shows that the "fused" corolla of many angiosperms may be produced by phylogenetically and developmentally different routes. I am not trying to suggest that development is uninformative in assessing similarity between structures, because this is palpably untrue (witness, for example, the excellent paper by Kaplan [1980] on the development of *Acacia* phyllodes). However, we should not necessarily expect morphological comparisons between strongly modified structures to be particularly informative in this respect (Stevens 1984c), and we may have to be careful not to compare organisms that are taxonomically very distant and morphologically very different from each other. Molecular data may play a crucial role here.

There is a further problem. The fact that one structure has a more complex development that includes the entire developmental sequence of another structure tells us nothing about the direction of the evolutionary change between the two. It could have gone in either direction, the simpler sequence being either primitive or derived via some kind of developmental truncation. If characters subject to such change are polarized according to outgroup analysis, simple developmental sequences that are the result of both derived and primitive pathways may be assessed as the same—both are classed as primitive. The same result will occur if sole reliance is placed on developmental complexity in a two-taxon system (cf. Nelson 1978, but see Voorzanger and van der Steen 1982, among others, for criticism of Nelson's position). Thus neither outgroup analysis nor the simple use of ontogenetic criteria may produce a pattern congruent with the process that generated it (cf. Stevens 1980a). However, not all characters may be affected by developmental changes (Fink 1982); examples are chromosome number and structural changes in chromosomes. Certain kinds of genomic and chemical characters may also be less affected than others by changes affecting morphogenetic aspects of development. Such characters might support a cladogram with a topology different from that supported by developmental characters, and this conflict will suggest that a reexamination of the characters on which the different trees were based is needed.

Discussion

The examination and comparison of independent suites of characters, or characters divided up in different ways, or the use of different group-forming algorithms, have been mentioned several times. They are some of the most powerful tools that taxonomists have for investigating the structure of their cladograms. Structures that do not readily change with additional data, that are robust, are the desiderata. New data that are derived from evidence that is as independent as possible from other data sets should be consilient with or concur with the favored hypothesis of phylogeny (Kluge 1983; see also Hull 1983). However, as has been emphasized above, choice of cladograms is not a simple matter (Moss 1983).

As I have mentioned several times, we should try to be sure that our methods of analysis allow us to generate at least the patterns that current evolutionary theory suggests are possible, and not to use methods that will lead to the recognition of only certain types of pattern. It is to this extent that findings in evolutionary biology should temper taxonomic methodology, so that the pattern generated by taxonomic analysis can best inform evolutionary generalizations. Basically, then, the relationship between the two is one of reciprocal illumination (Hull 1967). There is no necessary sequence of operations in the whole process.

Perhaps the discussion is caused in part by a confusion of levels. Simple inspection of cladograms will tell us little about causal factors of evolution, although correlation between patterns of characters in cladograms and other features may suggest such factors. Although some aspects of the speciation process (which is not a unitary process) seem adequately supported for us to modify pattern analysis to cope with them, yet whatever general theory of evolution we hold, many of the current ideas on speciation are neutral as regards pattern analysis.

The process of historical change in organisms has been largely mediated by changes in development in the broad sense (e.g., Sattler 1974), although there may be disagreement as to the importance of some kinds of developmental change. Botanists at first study details of the morphology of the adult organisms and base their assessments of relatedness on such studies. As more details of both the morphology and the development of organisms become known, the similarity relationships of structures change, and so the relationships that are based on them also change. But inasmuch as development is an intermediary in the action of process on organisms, we are caught up in process to the extent that development, used as a character or group of characters, is important in the detection of pattern, and to the extent that procedures are introduced to allow us to deal with the complexities and vagaries of development. There are areas in which evolutionary theory has been used in both a premature and basically circular fashion, for example, in evaluation of biogeographic pattern independent of phylogenetic relationships. Whatever we consider to be the simple methodological principles that we are using, we should be concerned about the detection of pattern using these principles if they run counter to reasonable theories as to how organisms have changed historically. Such principles should continually be reevaluated, since they may generate groups that cannot function in any historical theory, whether it be (neo-) Darwinian evolution or something else.

A comparison of historical reconstruction in linguistics, botany, and zoology reveals important and perhaps illuminating similarities as well as differences. Linguists talk about phonological laws of change that either are without exception or have exceptions that can reasonably be explained away (e.g., words that refer to taboo items changing more than words in more general use). Such laws in biology must be demonstrated or confirmed through his-

torically based analysis rather than assumed; they will emerge from the taxonomic patterns we produce or from a consideration of the physico-chemical
variables that affect development. Linguists mention that old forms of a language should be used in preference to current spoken/written forms because
they will reveal the relationships of that language most clearly. However, as
Crane/Hill and Novacek in this volume agree, in organisms fossils do not hold
a privileged status in phylogeny reconstruction.[1] In his chapter Cameron
shows how one can sometimes piece together many of the details of the
change of manuscripts over time. Plants with their relatively simple developmental pathways tend to present problems in phylogenetic reconstruction, and
they are like languages with no written records; more complex organisms with
complex developmental sequences can perhaps be compared with languages
that have a written record.

The issue of hybridization or reticulate evolution presents real problems
both to linguists, who apparently have tended to evade the issue (in part at
least for political and sociological reasons), and to botanists. To what extent
can those who study reticulating groups rely on a methodology initially devised for the detection of nonreticulate evolution? One solution might be to
advance to a level at which social or biological hybridization disappears, but
that ignores the immediate problem: "hybrid languages" or hybrid organisms
still have a history. The other approach is to modify methodology and to interpret branching patterns with extreme caution. William Wang has commented
(in discussion) that linguistic typology—a complex operation affected by
many things, including the history of the language—and historical analysis
are different operations and that in some contexts the former seems to be more
appropriate. This is apparently contradicted by Farris' assertion (1983) that
any goals of phenetic analysis are better subsumed and answered by cladistic
analysis. To a degree, Farris and Wang were talking about levels of knowledge
in different systems (see also Sokal 1983), as would also seem to be the case
in the historical analysis presented earlier. Furthermore, if in reconstruction
of the historical relationships of languages the pattern becomes clearer as you
move from the more recent past to the less recent past, this might also apply to
botanical patterns—hybridization at lower levels is confusing, but at somewhat higher taxonomic levels branching patterns may become clearer (although
I have suggested that this may not reflect actual historical relationships). Perhaps at higher levels in both language and botany, historical reconstruction
again becomes difficult. Evolution has occurred, but the level of our analysis
of the characters that have evolved and the methodology we used to construct
cladograms may sometimes suggest no particular phylogenetic pattern that is
interpretable as the result of this evolution.

Note

1. The scholars and their conclusions or comments referred to in this and the following paragraph were participants in the 1983 Interdisciplinary Round-Table (see Preface). Their contributions are included in this volume. I am very grateful to the members of the round-table for two days of stimulating discussion; to M. J. Donoghue, E. A. Kellogg, B. D. Mishler, S. E. Sultan, L. Wiener, and A. L. Weitzman for reading various drafts of the manuscript; and to R. A. Balan for typing all the drafts.

References

Adams, E. N. 1972. "Consensus Techniques and the Comparison of Taxonomic Trees." *Syst. Zool.* 21:390–397.

Almeida, M. T., and Bisby, F. A. 1984. "A Simple Method for Establishing Taxonomic Characters from Measurement Data." *Taxon* 33:405–409.

Ashton, P. S. 1982. "Dipterocarpaceae." In *Flora Malesiana,* series 1, vol. 9, part 2, ed. C. G. G. J. van Steenis. The Hague: Nijhoff.

Barabé, D. 1982. "Le critère de comparaison hors-groupe et son application à la systématique des Angiospermes: Cas de Hamamelidales." *C. R. Acad. Sci. Paris,* Sér. 3, 295:755–758.

Basile, D. V., and Basile, M. R. 1983. "Desuppression of Leaf Primordia of *Plagiochila arctica* (Hepaticae) by Ethylene Antagonists." *Science* 220:1051–1053.

Baum, B. R. 1983. "A Phylogenetic Analysis of the Tribe Triticeae (Poaceae) Based on Morphological Characters of the Genera." *Can. J. Bot.* 61:518–535.

Beatty, J., and Fink, W. L. 1979. *"Simplicity"* (review). *Syst. Zool.* 28:643–651.

Beckner, M. 1959. *The Biological Way of Thought.* New York: Columbia University Press.

Bessey, C. E. 1893. "Evolution and Classification." *Proc. Amer. Assoc. Adv. Sci.* 42:237–241.

Bolick, M. 1981. "A Cladistic Analysis of *Salmea* DC. (Compositae—Heliantheae)." In *Advances in Cladistics,* ed. V. A. Funk and D. R. Brooks, pp. 115–125. New York: New York Botanical Garden.

Brady, R. H. 1982. "Theoretical Issues and 'Pattern Cladistics.'" *Syst. Zool.* 3: 286–291.

———. 1983. "Parsimony, Hierarchy, and Biological Implications." In *Advances in Cladistics,* vol. 2, ed. N. A. Platnick and V. A. Funk, pp. 49–60. New York: Columbia University Press.

Bremer, K., and Wanntorp, H. E. 1978. "Phylogenetic Systematics in Botany." *Taxon* 27:317–329.

———. 1979. "Hierarchy and Reticulation in Systematics." *Syst. Zool.* 28:624–627.

Cannon, S. F. 1978. *Science in Culture: The Early Victorian Period.* New York: Dawson and Science History Publications.

Cantino, P. D. 1982. "Affinities of the Lamiales: A Cladistic Analysis." *Syst. Bot.* 7:237–248.

Cartmill, M. 1982. "Assessing Tarsier Affinities: Is Anatomical Description Phylogenetically Neutral?" *Geobios,* Mém. Spec., 6:279–287.

Coombs, E. A. K.; Donoghue, M. J.; and McGinley, R. J. 1981. "Characters, Computers and Cladograms: A Review of the Berkeley Cladistics Workshop." *Syst. Bot.* 6:359–372.

Crawford, D. J. 1985. "Electrophoretic Data and Plant Speciation." *Syst. Bot.* 10: 405–416.

Davis, P. H., and Heywood, V. H. 1963. *Principles of Angiosperm Taxonomy*. Edinburgh: Oliver & Boyd.

Donoghue, M. J. 1983a. "A Preliminary Analysis of Phylogenetic Relationships in *Viburnum* (Caprifoliaceae s. l.)." *Syst. Bot.* 8:45–58.

———. 1983b. "The Phylogenetic Relationships of *Viburnum*." In *Advances in Cladistics*, vol. 2, ed. N. I. Platnick and V. A. Funk, pp. 143–166. New York: Columbia University Press.

Ehrendorfer, F.; Krendl, F.; Habeler, E.; and Sauer, W. 1968. "Chromosome Numbers and Evolution in Primitive Angiosperms." *Taxon* 17:337–353.

Engler, A. 1892. *Syllabus der Vorlesungen über specielle und medicinisch-pharmaceutische Botanik*. Berlin: Borntraeger.

Farris, J. S. 1982. "Outgroups and Parsimony." *Syst. Zool.* 31:328–334.

———. 1983. "The Logical Basis of Phylogenetic Analysis." In *Advances in Cladistics*, vol. 2, ed. N. I. Platnick and V. A. Funk, pp. 7–36. New York: Columbia University Press.

Felsenstein, J. 1978a. "The Number of Evolutionary Trees." *Syst. Zool.* 27:27–33.

———. 1978b. "Cases in Which Parsimony and Compatibility Methods Will Be Positively Misleading." *Syst. Zool.* 27:401–410.

Fink, S. V. 1982. "Report of the Second Annual Meeting of the Willi Hennig Society." *Syst. Zool.* 31:180–197.

Funk, V. A.. 1981. "Special Concerns in Estimating Plant Phylogenies." In *Advances in Cladistics*, ed. V. A. Funk and D. R. Brooks, pp. 73–86. New York: New York Botanical Garden.

———. 1982. "The Systematics of *Montanoa* (Asteraceae: Heliantheae)." *Mem. New York Bot. Gard.* 36:1–133.

Funk, V. A., and Stuessy, T. F. 1978. "Cladistics for the Practicing Plant Taxonomist. *Syst. Bot.* 3:159–178.

Gosliner, T. M., and Ghiselin, M. T. 1984. "Parallel Evolution in Opisthobranch Molluscs and Its Implications for Phylogenetic Methodology." *Syst. Zool.* 33: 255–274.

Grant, V. 1981. *Plant Speciation*, 2d ed. New York: Columbia University Press.

Gray, A. 1842. *The Botanical Text Book*. New York: Wiley & Putnam.

Guédès, M. 1983. "Botanical Cladistics: Nothing New? A Response from Guédès." *Taxon* 32: 277–278.

Hart, J. A. 1983. "Systematic and Evolutionary Studies in the Genus *Lepechinia* (Lamiaceae)." Ph.D. thesis, Harvard University.

Heiser, C. B. 1973. "Introgression Reexamined." *Bot. Rev.* 39:347–366.

Hennipman, E., and Roos, M. C. 1982. *A Monograph of the Fern Genus Platycerium (Polypodiaceae)*. Amsterdam: North Holland.

Hooker, J. D. 1856. "On the Structure and Affinity of the Balanophoreae." *Trans. Linn. Soc. London* 22:1–68.

Hull, D. L. 1967. "Certainty and Circularity in Evolutionary Taxonomy." *Evolution* 21:174–189.

————. 1970. "Contemporary Systematic Philosophies." *Ann. Rev. Ecol. Syst.* 1:19–54.

————. 1979. "The Limits of Cladism." *Syst. Zool.* 28:416–440.

————. 1983. "Karl Popper and Plato's Metaphor." In *Advances in Cladistics,* vol. 2, ed. N. I. Platnick and V. A. Funk, pp. 177–189. New York: Columbia University Press.

Humphries, C. J. 1979. "A Revision of the Genus *Anacyclus* L. (Compositae: Anthemidae)." *Bull. Brit. Mus. Nat. Hist.,* Bot. Ser., 7:83–142.

————. 1983. "Primary Data in Hybrid Analysis." In *Advances in Cladistics,* vol. 2, ed. N. I. Platnick and V. A. Funk, pp. 89–103. New York: Columbia University Press.

Jepsen, G. L. 1944. "Phylogenetic Trees." *Trans. New York Acad. Sci.,* ser. 2, 6:81–92.

Judd, W. S. 1981. "A Monograph of *Lyonia* (Ericaceae)." *J. Arnold Arbor.* 62:63–209, 315–436.

Kaplan, D. 1980. "Heteroblastic Leaf Development in *Acacia:* Morphological and Morphogenetic Implication." *La Cellule* 73:135–203.

Kluge, A. 1983. Preface to *Advances in Cladistics,* vol. 2, ed. N. I. Platnick and V. A. Funk. New York: Columbia University Press.

Kluge, A., and Farris, J. S. 1969. "Quantitative Phyletics and the Evolution of Anurans." *Syst. Zool.* 18:1–32.

Knobloch, I. W. 1976. "Pteridophyte Hybrids." *Publ. Mus. Michigan State Univ.,* Biol. Ser., 5:277–352.

Kress, W. J., and Stone, D. E. 1983. "Morphology and Phylogenetic Significance of Exine-less Pollen of *Heliconia* (Heliconiaceae)." *Syst. Bot.* 8:149–187.

LaDuke, J. C. 1982. "Revision of *Tithonia.*" *Rhodora* 84:453–522.

Lenoir, T. 1978. "Generational Factors in the Origin of "Romantische Naturphilosophie." *J. Hist. Biol.* 11:57–100.

Levin, D. A. 1978. "The Origin of Isolating Mechanisms in Flowering Plants." *Evol. Biol.* 11:185–317.

Löve, À.; Löve, D.; and Pichi Sermolli, R. E. G. 1977. *Cytotaxonomic Atlas of the Pteridophyta.* Vaduz: Cramer.

Maddison, W. P.; Donoghue, M. J.; and Maddison, D. R. 1984. "Outgroup Analysis and Parsimony." *Syst. Zool.* 33:83–103.

Mayr, E. 1963. *Animal Species and Evolution.* Cambridge, Mass.: Belknap Press of Harvard University Press.

————. 1969. Principles of Systematic Zoology. New York: McGraw-Hill.

————. 1982. *The Growth of Biological Thought.* Cambridge, Mass.: Belknap Press of Harvard University Press.

Mishler, B. D., and Donoghue, M. J. 1982. "Species Concepts: A Case for Pluralisim." *Syst. Zool.* 31:491–503.

Moss, W. W. 1983. "Report on NATO Advanced Study Institute on Numerical Taxonomy." *Syst. Zool.* 32:76–83.

Nelson, G. J. 1978. "Ontogeny, Phylogeny, Paleontology, and the Biogenetic Law." *Syst. Zool.* 27:324–345.

————. 1983. "Reticulation in Cladograms." In *Advances in Cladistics,* vol. 2, ed. N. I. Platnick and V. A. Funk, pp. 105–111. New York: Columbia University Press.

Nelson, G., and Platnick, N. 1981. *Systematics and Biogeography.* New York: Columbia University Press.

Ospovat, D. 1981. *The Development of Darwin's Theory: Natural History, Natural Theology and Natural Selection.* Cambridge: Cambridge University Press.

Platnick, N. 1979. "Philosophy and the Transformation of Cladistics." *Syst. Zool.* 28:527–546.

Riggins, R., and Farris, J. S. 1983. "Cladistics and the Roots of the Angiosperms." *Syst. Bot.* 8:96–101.

Rosen, D. E. 1979. "Fishes from the Uplands and Intermontane Basins of Guatemala: Revisionary Studies and Comparative Biogeography." *Bull. Amer. Mus. Nat. Hist.* 162:267–376.

———. 1982. "Do Current Theories of Evolution Satisfy the Basic Requirements of Explanation?" *Syst. Zool.* 31:76–85.

Sanders, R. W. 1981. "Cladistic Analysis of *Agastache* (Lamaiceae)." In *Advances in Cladistics,* ed. V. A. Funk and D. R. Brooks, pp. 95–114. New York: New York Botanical Garden.

Sattler, R. 1974. "A New Approach to Gynoecial Morphology." *Phytomorphology* 24:22–34.

———. 1979. "'Fusion' and 'Continuity' in Floral Morphology." *Notes Roy. Bot. Garden Edinburgh* 36:397–405.

———. 1984. "Homology, a Continuing Challenge." *Syst. Bot.* 9:382–394.

Semple, J. C. 1981. "A Revision of the Goldenaster Genus *Chrysopsis* (Nutt.) Ell. Nom. Cons. (Compositae-Astereae)." *Rhodora* 83:323–384.

Simpson, G. G. 1961. *Principles of Animal Taxonomy.* New York: Columbia University Press.

Sneath, P. H. A., and Sokal, R. R. 1973. *Numerical Taxonomy,* 2d ed. San Francisco: W. H. Freeman.

Sober, E. R. 1983. "Parsimony Methods in Systematics." In *Advances in Cladistics,* vol. 2, ed. N. I. Platnick and V. A. Funk, pp. 37–47. New York: Columbia University Press.

Sokal, R. R. 1983. "A Phylogenetic Analysis of the Caminalcules, IV: Congruence and Character Stability." *Syst. Zool.* 32:259–275.

Sokal, R. R., and Sneath, P. H A. 1963. *Principles of Numerical Taxonomy.* San Francisco: W. H. Freeman.

Stevens, P. F. 1971. "A Classification of the Ericaceae, Subfamilies and Tribes." *Bot. J. Linn. Soc. London* 64:1–53.

———. 1980a. "Evolutionary Polarity of Character States." *Ann. Rev. Ecol. Syst.* 11:33–358.

———. 1980b. "A Revision of the Old World Species of *Calophyllum* (Guttiferae)." *J. Arnold Arbor.* 61:117–699.

———. 1981. "On Ends and Means, or How Polarity Criteria Can Be Assessed." *Syst. Bot.* 6:186–188.

———. 1984a. "Hauy and A.-P. de Candolle: Crystallography in the Development of Botanical Systematics and Comparative Morphology, 1780–1840. *J. Hist. Biol.* 17:69–82.

———. 1984b. "Metaphors and Typology in the Development of Botanical Systematics, 1690–1960, or the Art of Putting New Wine in Old Bottles." *Taxon* 33:169–211.

————. 1984c. "Homology and Phylogeny: Morphology and Systematics." *Syst. Bot.* 9:395–409.

————. 1986. "Evolutionary Classifications in Botany, 1960–1985," *J. Arnold Arbor.* 67:313–339.

Voorzanger, B., and van der Steen, W. J. 1982. "New Perspectives on the Biogenetic Law?" *Syst. Zool.* 31:20–205.

Wagner, W. H., Jr. 1952. "The Fern Genus *Diellia:* Its Structure, Affinities, and Taxonomy." *Univ. Calif. Publ. Bot.* 26:1–167.

————. 1980. "Origin and Philosophy of the Groundplan-Divergence Method of Cladistics." *Syst. Bot.* 5:173–193.

————. 1983. "The Recognition of Hybrids and Their Role in Cladistics and Classification." In *Advances in Cladistics,* vol. 2, ed. N. I. Platnick and V. A. Funk, pp. 63–79. New York: Columbia University Press.

Wanntorp, H. E. 1983. "Reticulated Cladograms and the Identification of Hybrid Taxa." In *Advances in Cladistics,* vol. 2, ed. N. I. Platnick and V. A. Funk, pp. 81–88. New York: Columbia University Press.

Watrous, L. E., and Wheeler, Q. D. 1981. "The Outgroup Comparison Method of Character Analysis." *Syst. Zool.* 30:1–11.

Whalen, M. B.; Costich, D. E.; and Heiser, C. B. 1981. "Taxonomy of *Solanum* Section *Lasiocarpa.*" *Gentes Herb.* 12:41–129.

Wheeler, Q. D. 1981. "The Ins and Outs of Character Analysis: A Response to Crisci and Stuessy." *Syst. Bot.* 6:297–306.

Wiley, E. O. 1981a. *Phylogenetics.* New York: John Wiley.

————. 1981b. "Convex Groups and Consistent Classifications." *Syst. Bot.* 6:346–358.

Young, D. A. 1981. "Are the Angiosperms Primitively Vesselless?" *Syst. Bot.,* 6:313–330.

————. 1983. "Botanical Cladistics: Nothing New?" *Taxon* 32:275–277.

9

Characters and Cladograms: Examples from Zoological Systematics

MICHAEL J. NOVACEK

The increasing popularity of Hennig's (1966) phylogenetic systematics, or cladistics, is a reflection of the satisfaction many proponents find in what they regard as a more rigorous and explicit methodology. Application of cladistics to particularly thorny systematic problems is not, however, always a cause célèbre. Some questions of relationships continue to elude cladists and noncladists, largely because of a poor understanding of what might be construed as "systematically meaningful" characters. The input of character evidence for cladistic analysis represents the gray area of the method. Characters may be variably recognized and defined by workers who follow the same basic cladistic procedure. There is no established formula for how many characters need to be considered in a "thorough analysis," no criterion for recognizing certain similarities as one, two, or many traits, no established dictum on the relative importance of one character over another. One may hope that one's analysis of characters is reasonably repeatable, but in practice cladists and noncladists alike will continue to argue over interpretations of characters, including seemingly simple issues involving the name and description of a trait. It seems that analogous problems arise in linguistics. The ways in which languages are described by certain traits have great influence on the ways language family trees are constructed. In this chapter, several problems pertaining to the assessment and use of characters are identified. Some of these problems are widely recognized, but reconsideration of them seems useful both in the context of this forum and in light of evolving methods in systematics.

Cladistic analysis is the search for hierarchical patterns that consist of nested sets of items. In biology, these items are taxa, of various levels of inclusiveness, whose recognition is based on distributions of biological characters. A particular hypothesis about a hierarchical pattern is rarely falsified in the strictest sense, that is, it is rarely abandoned because of a single contradictory character distribution. Under a parsimony approach, we accept the hypothesis that faces the fewest instances of contradiction (Platnick 1977). In this way, the preferred scheme requires the fewest supplementary, or ad hoc, hypotheses to explain the contradictions. In evaluating alternative cladograms, we must have some *a priori* knowledge, or at least assumptions, about the nature of the characters used in the analysis. Without such knowledge, contradictions in character distributions have at least three equally likely explanations. They may represent (1) characters independently acquired in the unrelated groups or (2) characters that represent the retention of a generalized, or plesiomorphic, condition for the taxa in question, or (3) characters that superficially appear similar but are not indicative of real resemblance (see also Platnick 1977; Gaffney 1979).

This ambiguity in explanation of contradictory characters does not operationally apply to a particular problem set of taxa. Cladistics is distinctive in using only specialized traits, or apomorphs, as evidence for groupings. Hence, the only possible contradictions to a particular cladistic hypothesis are apomorphs shared by two or more unrelated taxa, interpreted in an evolutionary context as convergence. Explanations 2 and 3 do not apply to such contradictions. This does not mean that our identifications of the apomorphs are correct. Such identifications are lower-level hypotheses that serve only as "observations" bearing on a particular cladistic hypothesis. As hypotheses, they are in themselves refutable (Platnick, in Harper and Platnick 1978).

Refutation seems most obvious in cases where any cited instance of resemblance is subject to explanation 3. If we are not confident that our perception of identity in structure is accurate, there seems to be little basis for further analysis. Suppose, for example, that the wings of bats and birds are cited as either a defining character of a group containing these taxa or as a feature independently acquired by these taxa. Further examination of the wings of these taxa may reveal (actually has revealed) that there is no evidence for special resemblance in the components of these systems. Since these components can be described in a more explicit fashion than the character-concept "wings," they provide a more effective index of resemblance. What is often recognized as a compelling case of resemblance—especially one that hinges on a "gestalt" similarity—may in reality be an artifact of a poor description.

Systematists will forever scrutinize each other when it comes to describing characters and identifying resemblance. One may be quick to defend one's own perception of character resemblance, but may be less confident about identification of such resemblance as apomorphic for a given taxonomic level.

Critics of cladistics claim the distinction of apomorphic from generalized traits to be the Achilles' heel of the method. For example, Harper (in Harper and Platnick 1978, p. 354) states that the recognition of specialized traits is based on "speculative criteria." It is clear, however, that since the seventeenth century many predictive classifications have been formulated through use of specialized traits to define groups. These traits were recognized by their restricted distribution within a set of taxa. In recent years, a more formal version of this procedure has been described under the label "outgroup comparison," wherein the generalized, or plesiomorphic, features are recognized as those present at a higher level of universality than the level of the group under investigation (the ingroup). For example, in most species of living mammals the young are born without any shell or shell-like covering. By contrast, the duck-billed platypus and the echidna (known collectively as the monotremes) are egg-layers. Which of these conditions is plesiomorphic? The nearest living outgroup of mammals is most likely some group of reptiles, and all but a few reptiles lay eggs. Other vertebrate outgroups such as amphibians and different groups of fish are also best described, despite some exceptions, as egg-layers. All birds are egg-layers. Hence there is overwhelming consensus that the egg-laying condition is plesiomorphic for these vertebrate groups and that monotremes simply retain this plesiomorphic condition (see Luckett 1977 for a recent review). Note that commonality of trait is given here with reference to a hierarchy. More than four thousand Recent species of mammals bear their young live, and only six Recent species lay eggs. It is the condition that best characterizes the outgroup and is shared by some members of the ingroup, not simply the most common condition in mammals, that can be identified as plesiomorphic.

Outgroup comparison does not preclude assessment of character distributions restricted to the ingroup. In such cases, it is easy to appreciate the role of parsimony in separating apomorphs from generalized traits (Farris 1982). Table 9.1 shows the distribution of alternative states of seven characters in mammals (monotremes and all groups to the right of monotremes) and some nonmammalian outgroups. Character states 1b (presence of mammary glands) and 2b (presence of hair) are useful as synapomorphies, that is, they represent shared apomorphs that define a monophyletic group, mammals. Character state 3b, as described above, defines the therians, or live-bearing mammals. Character states 4b and 5b are developmental traits unique within vertebrates to the placental, or eutherian, mammals. The trophoblast (5b) is a protective, extraembryonic layer less well differentiated in marsupials ("pouched" animals) and monotremes. The allantois (4b) is an extra embryonic "sac," which in most vertebrates serves as a reservoir for nutritive wastes and for gas exchange. In eutherians, the allantois has become greatly

Table 9.1. Distributions of States of Seven Characters in Selected Mammalian and Nonmammalian Groups (Characters are discussed in text)

1. (b)	Birds	Reptiles	Monotremes	Marsupials	Edentates	Insectivores	Bats	Macaque	Chimpanzee
1. (a) absence, (b) presence of mammary glands	a	a	b	b	b	b	b	b	b
2. (a) absence, (b) presence of hair	a	a	b	b	b	b	b	b	b
3. (a) Egg-laying, (b) live-bearing	a	a	a	b	b	b	b	b	b
4. (a) Large yolk-sac or yolk-sac placenta, (b) allantoic placenta	a	a	a	a	b	b	b	b	b
5. (a) absence, (b) presence of modified trophoblast	a	a	a	a	b	b	b	b	b
6. (a) columnar, (b) horseshoe-shaped stapes	a	a	a	a	a	b	b	b	b
7. (a) Incus small, dorsal to malleus, malleus articulated with ectotympanic, (b) Incus large, rotated, malleus detached from ectotympanic	—	—	a	a	a	a	a	b	b

expanded to contribute to an elaborate placenta for the prolonged development of the fetus within the maternal reproductive tract (see Lillegraven 1975; Luckett 1977). In marsupials, the placenta is formed from the yolk sac, and development within the mother is much shorter in duration (one marsupial group has a placenta formed from an allantois, but this structure is clearly nonhomologous with the eutherian placenta; see Luckett 1977). Character state 6b defines a subgroup of eutherian mammals in which the stapes is a horseshoe-shaped bone, sometimes referred to as a "stirrup" (some edentates and marsupials have a horseshoe-shaped stapes, but developmental and distributional evidence suggests that this is a secondary feature within these groups; see Novacek 1982).

Up to this point, reference to the nonmammalian outgroups is useful because one or the other state of each character is present in these outgroups. With character states 7a and 7b, however, we confront a different situation. The outgroups possess only one ear ossicle of the three, the stapes. Hence, these outgroups are irrelevant to a decision over which state of the malleus and incus (the "hammer" and "anvil" respectively) is plesiomorphic for mammals. Through a parsimonious consideration of congruent distributions in all the character states, we conclude that state 7a is plesiomorphic for mammals. In this condition, the incus is a small element, usually directly above and tightly articulated with the malleus. Moreover, the malleus is tightly articulated with the ectotympanic, the bony ring that embraces the ear drum. Such an ossicle condition provides an energy transformer of low mass and much stiffness of vibration. In contrast, monkeys and apes (and many other mammals not listed in Table 9.1) have ossicles with a large, rotated incus, and a malleus only loosely attached to the incus and detached from the ectotympanic (state 7b). This arrangement is characterized as the "freischwingender Typ" by Fleischer (1973) because it involves a sound transformer of low to moderate stiffnesss. State 7a is present and characteristic of each of the major mammalian groups (monotremes, marsupials, and eutherians), whereas state 7b is more restricted in distribution. Hence, we accept the pattern that best conforms to a hierarchy constructed from characters 1 through 6.

In this example, the scope of reference taxa for assessing characters shifts back and forth according to the assumed hierarchy and the character being evaluated. As Farris (1982) remarks, direct use of parsimony overcomes the limitations of certain prescriptions for outgroup comparison (e.g., Watrous and Wheeler 1981) by allowing for simultaneous rather than sequential evaluations of character-state distributions at different taxonomic levels. It should finally be noted that, although there is some latitude in the range or choice of the outgroup, the parsimony algorithm for assessment of character states can depend heavily on the states identified as plesiomorphic for the two nearest outgroups (Maddison, Donoghue, & Maddison 1984). Therefore, the strength of any character analysis will stand on the strengths of the higher-level branching sequence suggested by other characters.

Frequently, as in the foregoing example, characters are binary, that is, each character is represented by one generalized condition and one restricted condition. Characters may also be thought of as comprising states arranged sequentially (morphoclines) or arranged as diverging branches of a "character family." Morphoclines or other multistate schemes are, however, profoundly more refined hypotheses than polarity decisions on binary characters. There are several alternative sequences or branching pathways for any characters represented by several states. Some workers (e.g., Gaffney 1979) have discouraged the analysis of morphoclines as a general procedure. Many characters do, however, seem to be best described by a spectrum of modified conditions. Mickevich's (1982) transformation series analysis is a constructive approach to dealing with such multistate characters in a cladistic context. This procedure orders character states in a most parsimonious alignment with a cladogram of taxa based on distributions of less problematic (usually binary) characters. Cladistic analyses of such multistate characters are not strictly independent of *a priori* branching sequences. Nevertheless, as noted above, our judgment of states of all characters, whether binary or multistate, ultimately stems from certain assumptions of hierarchy for taxa and independent characters.

Given these assumptions, there has long been a desire to assess characters by direct means (for example, Nelson 1978), without reference to patterns suggested by other characters. Traditionally, such direct means for the identification of specialized states or complicated transformation sequences have emphasized the use of paleontological (fossil) and ontogenetic (developmental) data.

The fossil record was and still is regarded by many as the ultimate source of evidence for character and phylogenetic transformation. The recent wave of skepticism over the powers of fossil evidence arises from a logical impasse: discrepancies between fossils and other kinds of data can always be rescued by the *ad hoc* rationalization that the fossil record is incomplete (Schaeffer et al. 1972; Nelson 1978). Thus we can readily apply this explanation when data do not conform to such rules as "Older fossils show more primitive characters than younger fossils." Suppose that the fossil record shows earlier taxa to have a condition x and later taxa have condition x'. Applying the paleontological rule, we conclude that x is the primitive condition, and x' the more derived condition. Suppose we subsequently find a taxon with x' that occurs still earlier than taxa with x. From this discovery we may conclude that the paleontological rule is invalid as evidence for character transformation. We may also conclude, quite logically, that the paleontological rule is still generally valid, but in this particular case the fossil record was originally too incomplete to yield the correct pattern of change.

Thus if we wish to hold to the paleontological rule, we must ask *when* the fossil record is complete enough to provide reliable evidence of character transformation. We may have more confidence in a fossil record that conforms to patterns of character distribution derived purely from comparative biology without reference to a time scale. But the paleontological rule holds that the fossil record offers an independent line of evidence for character change, and hence that record cannot be judged purely in terms of its consistency with character polarities suggested by outgroup comparison or related approaches. Instead, one might attempt to measure the degree of completeness of a particular fossil record through geologic or stratigraphic analysis. Recent studies of this kind (Dingus and Sadler 1982; Schindel 1982) demonstrate that even some of the well-known examples of "excellent fossil records" share at least two critical flaws. First, the stratigraphic sequence containing the fossils is not a complete record. Different rates of sedimentation, removal of sediments through erosion, or physical deformation of sediments produce gaps in the rock sequence that represent gaps in time, thus missing populations or generations of organisms. Second, there is rarely satisfactory geographic control on a sequence of change represented by fossils. In other words, we rarely know whether the observed change holds for the one or few isolated rock sections where the fossils are sampled, for wider areas, or for the entire range of the fossils in question.

Given these difficulties, are there any fossil sequences that might be viewed as "reasonably complete"? The question seems impossible to answer in other than probabilistic terms. Through careful stratigraphic analysis, Schindel (1982) demonstrated that only a few well-known case studies deal with rock sections where the actual sediments represent more than 10 percent of the total temporal scope estimated for a given section. Moreover, only two of these studies could claim greater than 50 percent "completeness" with respect to sedimentation versus actual temporal span. These studies were ambitiously aimed at measuring the tempo and mode of speciation events between ancestors and descendants. Would these records, then, be unproblematic as evidence for a general pattern of character and phylogenetic transformation? A few depositional environments hold the most promise for such a goal. Unfortunately these environments—which are primarily limited to widespread sequences of ocean bottom deposits, shallow, offshore, marine sediments, and Late Cenozoic lake and stream deposits—have rarely been examined from both traditional paleo-stratigraphic and purely comparative (cladistic) approaches. One synthetic study (Miyazaki and Mickevich 1982) shows that inconsistencies between fossil occurrence data and branching sequences of morphologically based cladograms can be correlated, in some instances, with gaps in the stratigraphic record.

Unlike the fossil record, the ontogenetic record of an organism from fertilized egg to adult provides a more accessible and certainly more complete story. Of course, this completeness may be reduced by our own limits of observation and experimental design. These limitations, however, seem less difficult to overcome than in the case of a fossil record, where no refinement of analysis will fill the gaps that interrupt those historical records. Ontogenetic change has been widely endorsed by cladists and other systematists as a useful, independent line of evidence for character transformation. Yet there is a long-standing conflict over generalizations concerning the relationship between ontogeny and phylogeny, a conflict that can be traced back at least to the nineteenth century. Ernst Haeckel (1866) proposed, in a generalization now widely recognized as the biogenetic law, that "ontogeny recapitulates phylogeny." He maintained that the overall adult body plan of a taxon was the product of additions to the final stages of development in its ancestors. In this way, each ontogeny was succeeded by an extended ontogeny that allowed for appearance of new traits. To rescue the theory from absurdity, Haeckel further maintained that earlier stages of ontogeny were successively telescoped in the ontogenies of descendant taxa (see Løvtrup, 1978). The biogenetic law was in conflict with von Baer's concept of ontogeny. Von Baer (1866 and many earlier works) recognized the primitive or general nature of early developmental events and the specialized nature of late developmental events without drawing any particular phylogenetic implications from this observation. Adopting von Baer's concept in a phylogenetic context would lead one to claim that early stages of development of a descendant resemble the young stages, not the adults, of its ancestor. Recently, Nelson (1978) reframed a version of von Baer's "biogenetic law" aimed at elucidating character transformation. Nelson's rule stated that in the ontogenetic transformation from a more general character to a less general character, the more general character is primitive and the less general character is advanced.

How well do these theories of ontogeny explain character data? Many biologists disdain such theories because of the diversity of ontogenetic patterns of different rates and modes among related taxa. Such diversity—usually referred to as heterochrony—is claimed to provide too many exceptions to arguments linking ontogenetic pathways with decisions on polarity. Notable among the exceptions are cases of paedomorphosis (also called neoteny, although many embryologists regard neoteny as a special case of paedomorphosis). Paedomorphic descendants retain as adults some of the embryonic traits of their ancestors. The frequency of this pattern of "reverse capitulation" is unknown, although paedomorphosis has been cited as a major pattern in the phylogeny of salamanders and certain other animal groups.

That some biologists regard paedomorphosis as an exception to a more general pattern of recapitulation suggests that the latter ontogenetic pattern can provide some guide to character analysis. In detailed studies of ontogeny and phylogeny, paedomorphosis has been suggested but is usually dominated by trends that seem to demonstrate recapitulation (Miyazaki and Mickevich, 1982). Moreover, paedomorphosis implies that a portion of the original ontogeny was lost—hence, characters are lost. This can lead to ambiguity. Are the characters lost or merely transformed (see Nelson 1978)? The adult middle and inner ear of the echolocating bats at first appears to be simply the product of reverse recapitulation, involving the retardation of bone formation. This lends to the bony labyrinth and surrounding elements the appearance of the embryonic ear formed in cartilage. More detailed study reveals, however, that this retardation is only part of a mosaic of accelerated and arrested development that comprises an ontogeny unique to these animals. When these features are taken together, they represent an innovative step beyond the adult ear morphology present in the nearest relatives of echolocating bats (Novacek, manuscript in preparation). Many other putative cases of reverse recapitulation may actually be component trends of a mosaic and unique ontogeny of a specific set of characters.

Until such cases are examined on a broader scale, statements concerning the relative frequency of paedomorphosis or other heterochronic patterns seem simplistic and premature. At present, generalizations that equate earlier ontogenetic stages with more generalized or primitive conditions fail, as do many other hypotheses in biology, to provide inviolate theories of the natural world. Such generalizations may, however, be useful in guiding particular studies of character transformation, when the taxa concerned are known from a variety of defining characters and several ontogenetic patterns.

Thus far, we have considered only the problem of separating apomorphs from more generalized conditions, but there remains the problem of apomorphic characters whose distributions conflict with the most parsimonious cladogram. Such conflicts are usually interpreted as character homoplasy, as acquisition of independent traits through convergence or reversal. Indeed, many biologists are skeptical about the use of parsimony in unmasking patterns, because, they believe, evolution does not always work in parsimonious fashion, and homoplasy occurs with some significant frequency. Put another way, this argument is a claim that using the parsimonious tree to reveal homoplasy assumes that convergence does not occur most of the time, and this assumption is not justified, given the amount of homoplasy in many data sets.

This kind of argument gives rise to proposals for the *a priori* weighting of characters. Kirsch (1982), for example, suggests that biochemical characters (specifically, structural genes and their products), physiological processes, developmental programs, and gross anatomical characters comprise a scale of decreasing reliability in reconstructing phylogeny. Thus, biochemical characters would receive high scores, or weights, because they are characters not likely to mislead systematists. By contrast, tooth characters would receive low weights because of their homoplastic tendencies.

These arguments have logical problems. First, weighting of characters is not an alternative to parsimony. The distribution of weighted characters must, as in the case of unweighted characters, be evaluated by a parsimony procedure (Farris 1982). Second, to state that there is much homoplasy in large data sets is not the same as stating that homoplasy occurs with such rampant frequency that the signal (synapomorphy) is swamped out by the noise (homoplasy). If such were the case, convergence would not be recognized at all, because convergence can only be suspected within a framework of monophyletic groups defined by synapomorphies. Third, there is no compelling evidence that some characters are uniformally less prone to homoplasy than others. Arguments for the great reliability of biochemical characters, for instance, are based on theories of gene-biochemical interactions, levels and kinds of natural selection, and stochastic or constant rates of chemical evolution (e.g., the molecular clock models). These theories are themselves highly disputable. Fourth, attempts to weight characters with a formal scheme are arbitrary enough to lead to ambiguity over the ranking of characters. For example, should a form–function character complex be given higher weight than a character with developmental information? These issues are not clearcut, even when one adopts a weighting scheme.

In light of these points, most formal schemes for *a priori* weighting appear to be founded on dubious claims at best. It should, however, be noted that some form of character weighting, whether tacit or explicit, is probably part of every analysis. Our perceptions of characters are certainly influenced by some general theories regarding the relative significance of those characters. Moreover, some characters seem to warrant lower "weighting" simply on the basis of their lack of information content. In this category we might easily include "loss characters" for which there is little or no developmental information. Weighting of characters is, then, most likely inseparable from a process called character evaluation. But this kind of weighting is not a substitute for the parsimony analysis of distributions of explicitly defined characters.

When we reflect on the history of systematics, we are struck by the combination of insight, imagination, and coincidence that marks important advances in knowledge of relationships. A meaningful higher-level classification of mammals eluded workers of the seventeenth and eighteenth centuries, workers who emphasized the value of characters of the dentition and foot structure in the classifications of these organisms. It was not until the early nineteenth century that Blainville (1816), through his studies of reproductive anatomy, arrived at the basic higher-level phylogeny and classification of mammals that we still find acceptable. What led Blainville to this achievement? Was it a hunch that the reproductive system would provide a key for a meaningful classification? Or was it simply the conviction that this system should be studied "because it is there"? Whatever the reasons, they cannot be associated with an explicit methodology comparable to cladistic analysis. One of the most important contributions of cladistics is an emphasis on the distinction between specialized and generalized traits in a search for patterns of relationship. Such an emphasis has brought into focus the need for a more communicative understanding of characters used in phylogenetic reconstruction.

Acknowledgments

The author wishes to thank Bobb Schaeffer and John Maisey (American Museum of Natural History) and various participants of the Interdisciplinary Round-Table on Cladistics and Other Graph-Theoretical Representations for comments on the arguments presented.

References

Blainville, H. M. D. 1816. "Prodrome d'une nouvelle distribution systématique du règne animal." *Bull. Soc. Philom.* 1816:105–124.

Dingus, L., and Sadler, P. M. 1982. "The Effects of Stratigraphic Completeness on Estimates of Evolutionary Rate." *Syst. Zool.* 31:400–412.

Farris, J. S. 1982. "Outgroups and Parsimony." *Syst. Zool.* 31:328–334.

Fleischer, G. 1973. "Studien am Skelett des Gehörorgans der Säugetiere, einschliesslich des Menschen." *Säugetierkundl. Mitteilungen* (München), 21:131–239.

Gaffney, E. S. 1979. "An Introduction to the Logic of Phylogenetic Reconstruction." In *Phylogenetic Analysis and Paleontology*, ed. J. Cracraft and N. Eldredge, pp. 79–111. New York: Columbia University Press.

Haeckel, E. 1866. *Generelle Morphologie der Organismen, allgemeine Grundzüge der organischen Formen-Wissenschaft, mechanisch begründet durch die von Charles Darwin reformierte Descendenz-Theorie*, vols. 1 and 2. Berlin: George Reimer.

Harper, C. W., Jr., and Platnick, N. I. 1978. "Phylogenetic and Cladistic Hypotheses: A Debate." *Syst. Zool.* 27:354–362.

Hennig, W. 1966. *Phylogenetic Systematics*. Urbana: University of Illinois Press.

Kirsch, J. A. 1982. "The Builder and the Bricks: Notes Toward a Philosophy of Characters." In *Carnivorous Marsupials*, ed. M. Archer, pp. 587–594. Royal Zoological Society of New South Wales. Vol. 2.

Lillegraven, J. A. 1975. "Biological Considerations of the Marsupial-Placental Dichotomy." *Evolution* 29:707–722.

Løvtrup, S. 1978. "On von Baerian and Haeckelian Recapitulation." *Syst. Zool.* 27:348–352.

Luckett, W. P. 1977. "Ontogeny of the Amniote Fetal Membranes and Their Application to Phylogeny." In *Major Patterns in Vertebrate Phylogeny*, ed. M. K. Hecht, P. C. Goody, and B. M. Hecht, pp. 439–516. New York: Plenum Press.

Maddison, W. P.; Donoghue, M. J.; and Maddison, D. R. 1984. "Outgroup Analysis and Parsimony." *Syst. Zool.* 33:83–103.

Mickevich, M. F. 1982. "Transformation Series Analysis." *Syst. Zool.* 31:461–478.

Miyazaki, J. M., and Mickevich, M. F. 1982. "Evolution of *Chesapecten* (Mollusca: Bivalvia, Miocene-Pliocene) and the biogenetic law." *Evol. Biol.* 15:369–409.

Nelson, G. J. 1978. "Ontogeny, Phylogeny, Paleontology, and the Biogenetic Law." *Syst. Zool.* 27:324–345.

Novacek, M. J. 1982. "Information for Molecular Studies from Anatomical and Fossil Evidence on Higher Eutherian Phylogeny." In *Macromolecular Sequences in Systematic and Evolutionary Biology*, ed. M. Goodman, pp. 3–41. New York and London: Plenum Press.

Platnick, N. I. 1977. "The Hypochiloid Spiders: A Cladistic Analysis, with Notes on the Atypoidea (Arachnida, Araneae)." *Amer. Mus. Novitates* 2627:1–23.

Schaeffer, B.; Hecht, M. K.; and Eldredge, N. 1972. "Paleontology and Phylogeny." *Evol. Biol.* 6:31–46.

Schindel, D. E. 1982. "Resolution Analysis: A New Approach to Gaps in the Fossil Record." *Paleobiology* 8:340–353.

Von Baer, K. E. 1866. "Über Prof. Nic. Wagners Entdeckung von Larven, die sich fortpflanzen, Herren Garrens Verwandte und ergänzende Beobachtung und über die Pädogenesis überhaupt." *Bull. Acad. Imp. Sciences, St. Petersbourg* 9:63–137.

Watrous, L. E., and Wheeler, Q. D. 1981. "The Out-group Comparison Method of Character Analysis." *Syst. Zool.* 30:1–11.

10
Reconstructing Genetic and Linguistic Trees: Phenetic and Cladistic Approaches

MARYELLEN RUVOLO

> *The work of classification entails a conceptual model shared by all of anthropology. Underlying it is the fundamental question: How are resemblances and differences among peoples (their cultural, racial, linguistic characteristics) to be interpreted? The answers ultimately become part of a general theory of the nature and dynamics of human life, but the first task is sorting and mapping, ordering the universe with which one is dealing.*
>
> —Dell Hymes (1964)

For many biologists, phylogenetic reconstruction is the necessary first step in the search for understanding general biological processes. After all, how can we attempt to study the tempo and mode of evolution, for example, without reference to particular species groups for which we have believable evolutionary trees? Do certain constellations of traits evolve more quickly than others? When did particular species diverge from each other? We cannot begin to answer these questions without establishing the evolutionary branching patterns linking taxa we wish to study. As a subject in its own right, phylogenetics challenges us to rediscover the sequence of evolutionary events (which we know occurred only once) from the distribution of characters among taxa.

From the biological viewpoint at least, the parallel goals of historical linguistics and of phylogenetics are striking, and it seems reasonable that sharing analytical methods would prove fruitful. As Hymes states in the quotation at the beginning of this chapter, "the first task is sorting and mapping," knowing how the languages or dialects one studies are related. This is probably a more difficult job for linguists than for biologists (or at least the ones who don't study plants), because of the existence of hybrid forms. On the other hand, linguists reconstructing language relationships may have the advantage in that the preserved record of historical change is more dense. There are more written documents spread over a linguistically significant unit of time than there are fossils spread over an evolutionarily significant time period. Another linguistic goal is the establishment of time depths, or what biologists would refer to as divergence times, for languages (Gudschinsky 1964; Swadesh 1955, 1964). Almost all the problems associated with use of the "molecular clock" for establishing species divergence times also apply to the lexicostatistical/glottochronological approach to calculating time depths. The departure point for all these questions, whether in systematics or historical linguistics, consists of tree construction.

This chapter shows how evolutionary trees are constructed from one type of biological data and how these techniques (and the viewpoints underlying them) might be carried over into a linguistic framework. We will also see how the basic philosophical dichotomy between the two approaches to systematic biology, the phenetic and the "cladistic," can result in very different evolutionary tree reconstructions. The essential divergence point between these two schools lies in their treatment of similarity between groups of organisms.

Types and Uses of Similarity

Phenetic and cladistic analyses differ in their classifying and handling of similarities between taxa. Prior to Hennig (1966), biologists thought of similarities in a dyadic way. A similarity between two taxa was either "real" because it reflected shared inheritance, or was not real because it represented (coincidentally) the same, convergent solutions to some evolutionary challenge. Hennig broke up this dyadic scheme into a triadic one by splitting the category of "real" or inherited similarities. He distinguished between taxa resembling each other (1) because they inherit a trait from a remote common ancestor and (2) because they share a newly evolved trait unique to them and their immediate common ancestor. Hennig's insight was that only the latter category, the inherited, shared, more recently evolved traits (synapomorphies) between taxa, is useful for revealing evolutionary relatedness. For a cladistic analysis, therefore, only the patterns of shared derived traits are used to assess phylogenetic closeness; shared, primitive traits (symplesiomorphies), although inherited, are uninformative and so are convergent features (Figure 10.1). In contrast, a phenetic analysis draws no distinction between classes of similarities. All similarities form the basis for judging relatedness and are in this sense equivalent. It is this crucial difference in how one uses similarities to judge phylogenetic relatedness that differentiates the cladistic and phenetic viewpoints.

From a quick reading of the literature, it seems that historical linguistics has used a largely cladistic approach, in which types of similarities have been distinguished. In his editorial introductions to *Language in Culture and Society,* Hymes outlines four types of resemblances:

1. "generic," . . . "inherent in all units within the frame of analysis";
2. "convergent," . . . "due to chance, because of the limits of possible divergence, or some recurrent positive tendency such as sound symbolism or a functional correlation of traits";
3. "genetic," . . . "continuously transmitted to the units in question from a common ancestor";
4. "diffusional," . . . "transmitted from one unit to another subsequent to the period of any common ancestor."

The last two types of resemblances, genetic and diffusional, are "due to historical connection," unlike generic and convergent similarities. Of these two types of historical resemblances, the genetic ones are "fundamental to other lines of historical work" (Hymes 1964, p. 568). Language patterns changing through time which produce these various resemblances are sketched in Figure 10.2.

Diffusional and convergent resemblances produce the same patterns of character traits, making the two difficult to distinguish. As Sapir and Swadesh both recognized (Swadesh 1964), the "morphological kernel" or "archaic residue" of a language, if it is at all possible to delineate, is useful precisely because it eliminates the borrowed or "diffused" elements.

A Biological Example

To demonstrate the differences between phenetic and cladistic tree reconstruction, we will examine a biological example, starting in both cases with the same data set. This example focuses on eighteen closely related African

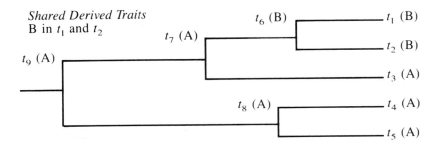

Shared Derived Traits
B in t_1 and t_2

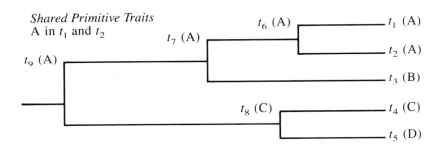

Shared Primitive Traits
A in t_1 and t_2

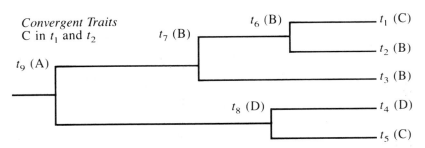

Convergent Traits
C in t_1 and t_2

Figure 10.1. Types of biological similarities. For biological taxa t_1 through t_9 known to be related as indicated below, we will consider one character with alternative traits represented as capital letters. For example, t_1 (B) means that taxon 1 expresses trait B.

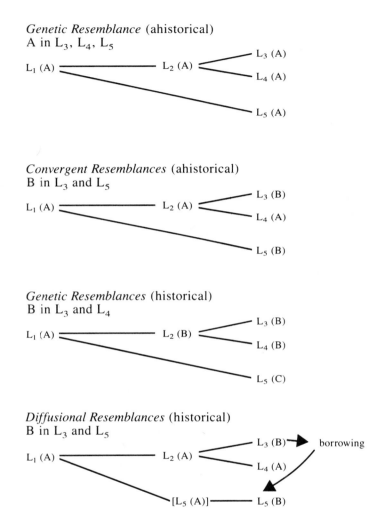

Genetic Resemblance (ahistorical)
A in L₃, L₄, L₅

L₁ (A) ————————— L₂ (A) ⟨ L₃ (A)
                                         L₄ (A)

                                         L₅ (A)

Convergent Resemblances (ahistorical)
B in L₃ and L₅

L₁ (A) ————————— L₂ (A) ⟨ L₃ (B)
                                         L₄ (A)

                                         L₅ (B)

Genetic Resemblances (historical)
B in L₃ and L₄

L₁ (A) ————————— L₂ (B) ⟨ L₃ (B)
                                         L₄ (B)

                                         L₅ (C)

Diffusional Resemblances (historical)
B in L₃ and L₅

L₁ (A) ————————— L₂ (A) ⟨ L₃ (B) ➤ borrowing
                                         L₄ (A)

                           [L₅ (A)] ——— L₅ (B)

Figure 10.2. Types of linguistic resemblances (after Hymes 1964). Ancestral languages L₁ and L₂ and their descendant forms L₃, L₄, and L₅ have alternative traits for a given linguistic feature, given by capital letters.

monkey species known as the Cercopithecini. Their close relationship is revealed by morphology, behavior, and distribution in the wild, but the molecular systematic relationships within the group had not been worked out until recently (Ruvolo 1983).

For each species, the biochemical traits were measured by electrophoresis of blood proteins, which involves the separation of proteins based on their size and overall electrical charge. The protein mixtures to be electrophoresed are inserted into slots of a jelly-like substance, known as a gel. Electric current applied along the length of the gel forces the proteins through the substance, the more negatively charged and/or the smaller ones moving faster than the positively charged and/or larger ones. When the electric field is removed, the different proteins will have migrated to various distances down the length of the gel. When a stain appropriate for the protein one wishes to visualize is applied, one or more colored bands appear in the places of protein concentration. The relative distance traveled down the length of the gel, also known as the relative mobility, is the primary set of data we observe. By other analyses, it can be shown that different relative mobilities reflect alternative forms of a gene, known as alleles, at a genetic locus. In essence, electrophoretic data are like any other data used for systematic purposes, in that for a given character (in this case, a genetic locus) alternative forms of the trait (in this case, alleles) exist and can be scored for individual organisms. What is unique to this type of data, unlike most morphological or behavioral data, is that one knows that the alleles are direct reflections of particular, single genes.

In all, fourteen genetic loci were scored, and the resultant data set was analyzed in several ways, but the basic distinction between cladistic and phenetic treatments is the key. Figure 10.3 sketches these approaches.

The phenetic analysis uses the frequencies of the alleles within each species to calculate a "distance" between each pair of species, summarized in matrix form. These observed distances are referred to as phenetic distances. A clustering algorithm converts the pairwise distance measurements into a branching network or phenogram. Each branch on the phenogram has an associated branch length. The patristic distance between any two species is defined as the sum of the branch lengths that must be traveled to connect the two species. Note that the patristic distance between two species can be (and usually is) different from the phenetic distance with which we started. In general, the goal is to construct a phenogram so that the differences between the patristic and phenetic distances, or homoplasies, summed over the entire tree are minimized. Another way of looking at the situation is that the observed (phenetic) distances must be stretched or shrunk slightly (yielding patristic distances) in order to summarize the data in the form of a tree-like network, but the goal is to keep such distortion to a minimum.

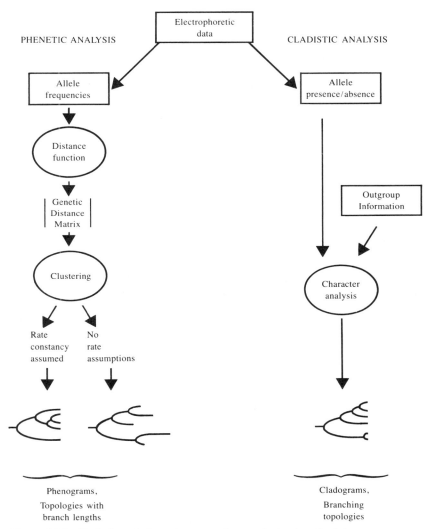

Figure 10.3. Cladistic and phenetic approaches to electrophoretic data.

Features of the phenetic approach which should be observed here are, first, that using all genetic loci equally is synonymous with treating all types of similarity equivalently, the hallmark of phenetics. The second point is that genetic distance functions are not all equivalent, in the sense that they preserve the same data structure or are scalar multiples of each other. Specifically, some genetic distance functions are metric, that is, the triangle inequality holds for them,[1] while others are not. Only the use of metric distance functions results in a phenogram with branch lengths interpretable as amounts of evolutionary change (Farris 1981). It is misleading even to refer to nonmetric functions as "distances," because they do not possess the requisite mathematical property. Unfortunately, one of the first and currently most widely used genetic "distance" functions (due to Nei 1972, 1978) is nonmetric (Farris 1981). In this study the Manhattan distance, a bona fide metric function, has been used (see Farris 1981 for elaboration).

The third noteworthy point is that some clustering algorithms implicitly assume constant rates of change, while others do not. It is surprising that this vital distinction has not been acknowledged by advocates of the molecular clock, and it still remains an issue of contention. The modified-distance Wagner technique, which does not assume rate constancy, is one of the clustering methods used here. (For development and discussion of this method, see Farris 1970, 1972, 1981.) Generally, it seems to be a good idea not to assume rate constancy. If rates are constant for a given data set, that will be revealed rather than assumed. Clustering techniques assuming rate constancy produce trees with the tell-tale property that the branch ends all line up neatly when the tree is drawn to scale.

Application of these phenetic techniques to the Old World monkey data set is shown in Figures 10.4, 10.5, and 10.6. These techniques differ in choice of distance measures and clustering algorithms. The effect of assuming constant genetic rate change or not is demonstrated by comparing Figures 10.4 and 10.5. The seductive simplicity of the first tree (Figure 10.4), which is the one traditionally drawn by genetic researchers, contrasts sharply with the messy picture produced when homogeneous rates are not assumed (Figure 10.5). Both phenograms, however, are based on Nei's nonmetric "distance" measure and are therefore not interpretable as summarizing evolutionary change. Using a properly metric distance measure offers us a phenogram (Figure 10.6) with all the desired qualities. The clustering technique (distance Wagner algorithm) produces an unrooted tree, which was rooted in this case by bending in half the branch leading to the species *(E. patas)* considered by primate systematists to be most morphologically distinct from the others.

In the cladistic analysis, the presence or absence of alleles rather than their frequencies is important. Evaluating the alleles as primitive or derived is

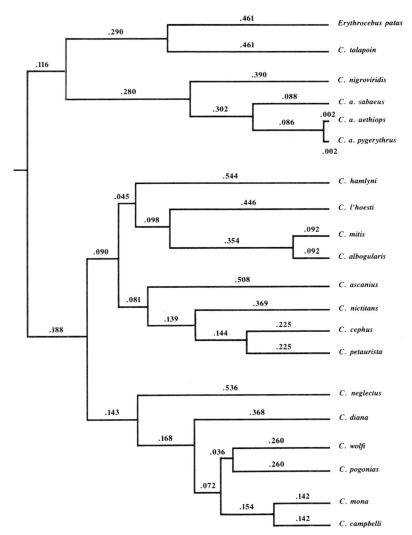

Figure 10.4. A phenogram of the Cercopithecini based on Nei's standard genetic distance and UPGMA clustering.

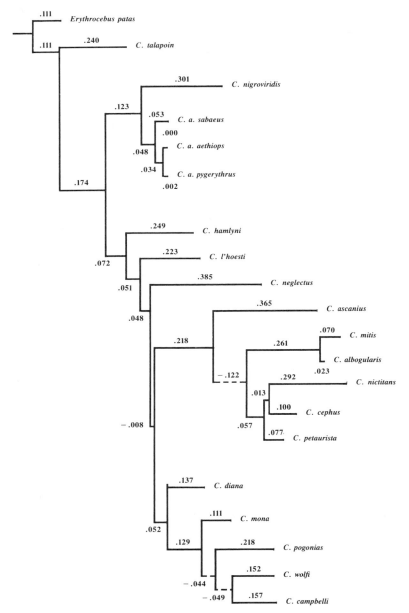

Figure 10.5. A phenogram of the Cercopithecini based on Nei's standard genetic distance and the distance Wagner method.

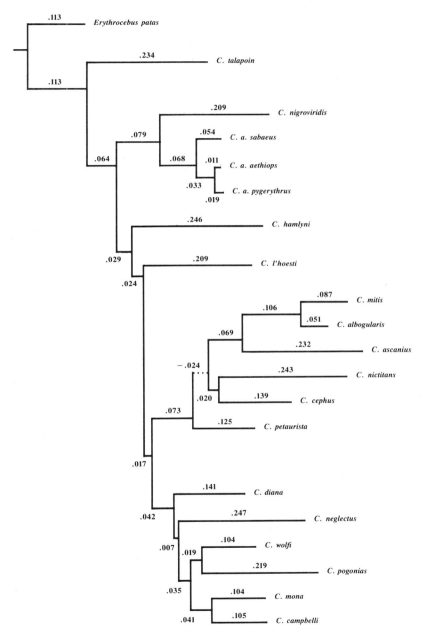

Figure 10.6. A phenogram of the Cercopithecini based on the Manhattan distance and the distance Wagner method.

the next step. To this end, a modified outgroup comparison was made, with a sister group of species, the Papionines. First, alleles present in the outgroup species were scored, and this allele list was compared with that of three in-group species *(nigroviridis, patas, talapoin)* thought to be more primitive on the basis of chromosomal and morphological evidence. The intersection of these two lists was used as the slate of primitive characters. This procedure (first used by Baverstock et al. [1979] for electrophoretic data) allows outside evidence to be brought in for a more accurate assessment of primitive traits. (See Wiley 1981 for further discussion of outgroup comparisons.)

The next step is character analysis, the discovery of nested sets of shared derived traits (synapomorphies) among the species we are studying. For any given character, in this case a genetic locus, we can draw a cladogram depicting a pattern of shared derived alleles. Not all loci will indicate the same cladogram, however. The aim is to come up with a cladogram that is most compatible with or most parsimonious for the allele patterns for all the loci taken as a whole. Inevitably one (or several) characters will show contradictions. Minimization of character contradictions leads to a most parsimonious cladogram. Each node of the cladogram will be defined by at least one particular shared derived trait, in this example, one allele of a genetic locus. Unlike a phenogram, a cladogram has no associated branch lengths; it is just a branching topology.

One cladogram constructed for the electrophoretic data set is shown in Figure 10.7. (It may or may not be most parsimonious, since it was constructed by hand and all other possibilities were not checked.) Indicated for each node is the abbreviated genetic locus name followed by a letter for a particular allele (e.g., "ADA c" means allele c at the ADA locus; upper-case P stands for the "primitive" allele). Here we can concretely observe the unique way in which similarities are handled cladistically. Not all genetic loci are treated equivalently; for any given node, some loci define it, whereas others are not important. Notice that in two cases rederivation of the primitive allele is indicated, suggesting a back-mutation (for ADA in *C. l'hoesti* and *C. hamlyni* and for DIA in *C. nictitans* and *C. mitis*). A minimal number of back-mutations characterizes a best-fitting cladogram.

How does this cladogram compare with the "best" phenogram we have constructed? To make comparison easier, we can reroot the original Wagner network of our "best" phenogram (Figure 10.6) to make it most like the cladogram. (This is legitimate because the network was originally unrooted, and we chose to bend it along one branch.) Since the first divergent species in the cladogram which cannot be rerooted is *nigroviridis,* we will use it to form the base of the phenetic network. To visualize the rerooting process, imagine

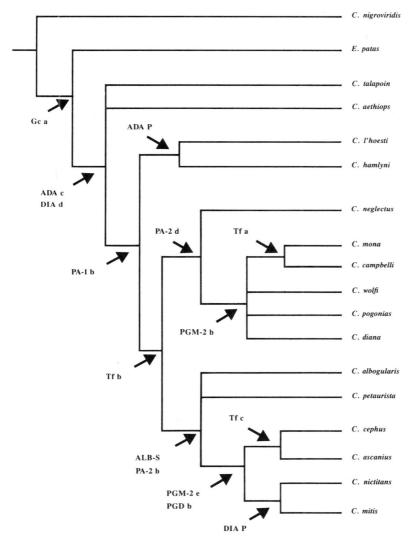

Figure 10.7. A cladogram of the Cercopithecini based on outgroup comparisons and character analysis. Derived alleles are indicated for each node.

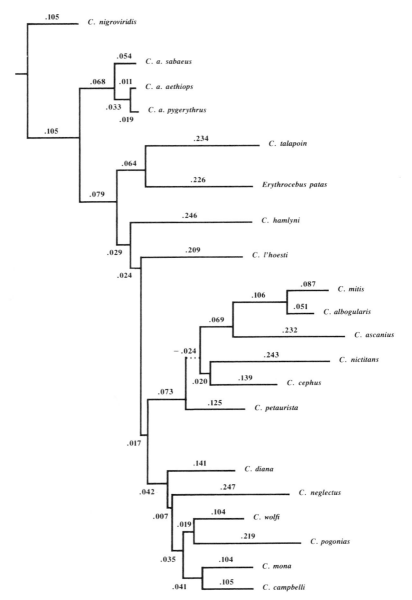

Figure 10.8. Rerooted version of Figure 10.6.

that the phenetic network (Figure 10.6) is built out of string and can be picked up in one spot (in this case halfway along the *nigroviridis* branch) and then set down again. None of the internal connections is changed. What results is shown in Figure 10.8.

Broadly, both phenogram and cladogram indicate the existence of the same two relatively derived species groups, in terms of membership although not in branching pattern (referred to here as the *"mitis"* and *"diana"* groups). Actually the cladogram leaves some internal relationships unresolved; more shared derived traits are needed to sort them out. Generally speaking, it is around the base of the trees where phenogram and cladogram differ most. (After *nigroviridis*, the next species to diverge is *aethiops* on the phenogram, but *patas* on the cladogram. *Talapoin* and *patas* are genetically close on the phenogram, but *talapoin* and *aethiops* are closer on the cladogram. *Hamlyni* and *l'hoesti* diverge sequentially in the phenogram but together on the cladogram.)

The above observation is probably a particular instance of a general phenomenon, namely, that cladistic and phenetic analyses of a data set tend to be more congruent for relatively derived species but less congruent for the relatively primitive ones. This is not surprising, considering their differing treatments of similarity types. Since phenetic analyses use all similarities, including shared primitive ones, indiscriminately, while cladistic analyses ignore shared primitive traits to concentrate on shared derived ones, any taxon that is relatively primitive may be discrepantly classified.

A Linguistic Example

The same analytical scheme that has been demonstrated here for biological data can be carried over into historical linguistics (Figure 10.9). Using a linguistic data set, the linguistic traits of interest would be directly analogous to genetic loci, because they are the variables we are studying. The particular traits scored for each language would correspond to genetic alleles, because they are the values assumed by the variables. Notice that the phenetic method using a clustering technique assuming constant rates of change is the classical lexicostatistical/glottochronological approach.

The following linguistic example (hesitantly presented by a nonlinguist) illustrates these approaches. This example is somewhat unusual in that it is typically phonological data (along with some morphological and syntactical data) that are analyzed cladistically by historical linguists, but herein is presented lexical data analyzed cladistically. Suppose we have languages L_1, L_2,

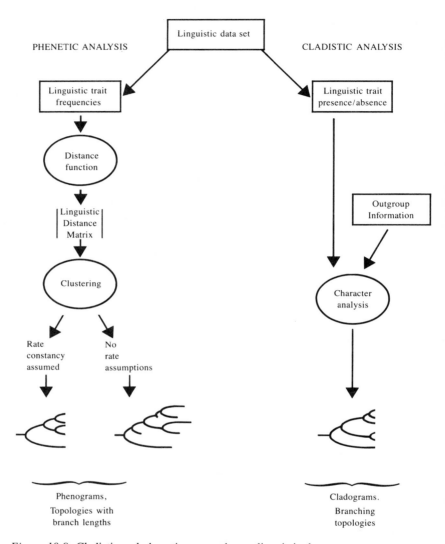

Figure 10.9. Cladistic and phenetic approaches to linguistic data.

and L_3, known to be related, and an outgroup language L_0 more distantly re-lated to them. How exactly are L_1, L_2, and L_3 related? Using a set of linguistic traits, in this case a core vocabulary of ten core definitions (for which borrow-ing is unlikely), we can score core words in L_1, L_2, L_3, and L_0 (given in Table 10.1). Beginning with the phenetic method, we can assign distances between languages X and Y as:

$$Distance(X,Y) = 1 - Similarity(X,Y),$$

where similarity is measured as the proportion of shared core words (cog-nates) between X and Y. The matrix below summarizes the similarity measure-ments in the upper half and distance measurements in the lower half.

	L_1	L_2	L_3	L_0	
L_1	—	.8	.5	.2	
L_2	.2	—	.6	.3	Similarity
L_3	.5	.4	—	.7	
L_0	.8	.7	.3	—	

Distance

Table 10.1   Ten-word core vocabulary in languages L_1, L_2, L_3, and outgroup lan-guage L_0. Each language is scored for the core words (lower-case letters) used for the core definitions. A core word shared by two or more languages is called a cognate word in the text.

Core Definitions	Core Words			
	L_1	L_2	L_3	L_0
1. "bird"	b	b	b	a
2. "heart"	d	d	c	c
3. "rock"	g	f	e	e
4. "tooth"	h	h	i	i
5. "night"	j	j	j	k
6. "mother"	l	l	l	l
7. "skin"	n	m	m	m
8. "father"	o	o	o	o
9. "rain"	p	p	p	q
10. "hunt"	s	s	r	r

Applying the unweighted-pair groups method of clustering (UPGMA), which assumes constant rate change, yields the phenogram in Figure 10.10. The distance Wagner algorithm shows a very different picture (Figure 10.11). (Rooting of the Wagner network was chosen along the L_0 branch because we know that it is an outgroup. It can also be rooted along any other branch; this does not alter any of the lengths.)

Notice that the rates of change are very unconstant in the Wagner tree. In particular, L_1 has changed three times as much as L_2. Assuming that these are contemporary languages, one would be hard pressed to assign time depths to them.

To construct a cladogram, the special significance of L_0 as the outgroup language is employed. Specifically, any core word present in L_0 will be scored as "primitive." Since there is only one available outgroup, this is a reasonable procedure. A cladogram for this data set is shown in Figure 10.12. The nodes are defined by the sharing of particular core words known to be derived (because they are not present in L_0). By referring back to the original table of core words, we can observe (1) that some core definitions ("father," "mother") are uninformative because all languages use the same core word and (2) that some core definitions ("rock," "skin") are uninformative because their derived core words are not shared by two languages.

This example does not illustrate character contradiction. If we had scored L_1 as having core word k for "night," for example, that would be an instance of contradiction because it suggests that L_1 is more closely related to L_0. Most likely our choice of the above cladogram as most parsimonious would not be changed by this one contradiction. The cladogram would be slightly less parsimonious but still the best fit to the data.

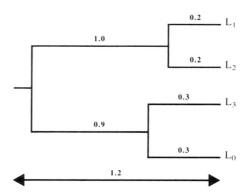

Figure 10.10. A linguistic phenogram using the UPGMA clustering technique.

The difference between the linguistic phenogram and the cladogram involves the relatively primitive languages L_3 and L_0. The phenetic distance measure sees them as close together because they share many primitive core words. Essentially this is the same discrepancy we observed in the biological example. The cladistic analysis is sensitive only to those few core words linking L_3 with L_1 and L_2, even though, overall, L_3 is most similar to L_0.

One data set can give us very different pictures of linguistic relationships according to the analytical method used. Which tree should we believe? As a practical matter, it seems productive to analyze data by all the various phenetic and cladistic techniques. In that way, if both phenetic and cladistic techniques indicate the same results for a data subset (perhaps for the more derived taxa of languages), then one can feel reasonably certain about that part of the tree. The choice between phenetic and cladistic trees is a larger issue resting ultimately on one's beliefs about the types and uses of similarity for revealing phylogenetic relationships.

Comments on Lexicostatistics/Glottochronology

As a biologist and outside observer looking in on lexicostatistical methodology, which shares the approach (and therefore the weaknesses) of phenetic analysis, I have the following comments.

The measures used to summarize similarities between languages could be borrowed from biology, particularly because they are sensitive to variability within the species (or languages) being compared. For example, Gudschinsky's (1964) summary advocates that when two or more cognates are used equally in a language, the best way to choose one (in order to score a language for that trait) is "by flipping a coin if necessary" (Gudschinsky 1964, p. 614). How-

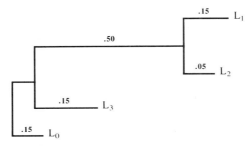

Figure 10.11. A linguistic phenogram using the distance Wagner method.

ever, a measure that reflects the language's diversity, for instance, one analogous to the genetic distance measures that capture the structure of allelic
distributions, might prove more accurate. After all, if two languages X and Y
each use cognates a and b 50% of the time, the coin-toss method leads to
different scorings for X and Y in half of all analyses, although both languages
use the two cognates with identical frequencies. In addition, when the cognate
distribution is

Languages

		X	Y
Cognates	a	.8	.4
	b	.2	.6

instead of scoring X and Y differently such a measure would incorporate the
fact that the two overlap 60 percent of the time (40 percent for a and 20 percent for b).

Glottochronologists should try to avoid tree-building techniques that
assume a constant rate of linguistic change. Rather than assuming rate constancy in a linguistic data set, it is preferable to demonstrate it (if true). Linguists could borrow the distance Wagner method of tree-building, for example,
from biological pheneticists. Instead of using pairwise shared cognate per-
centages for use with the time depth formula,[2] one could measure shared cognates among several languages, calculate distances between pairs, and apply
the distance Wagner method. This procedure would test whether word retention rates are constant, rather than assuming so.

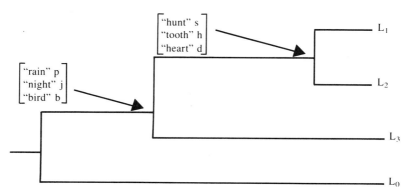

Figure 10.12. A cladogram of the linguistic data example.

For example, using Swadesh's (1964) French-English-German comparison and defining distance to be one minus the proportion of shared residual vocabulary, we have:

Distance (= 1 − Residual)

English–German	.40
English–French	.73
French–German	.71

Comparing English and German alone and assuming constant rates gives

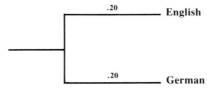

where the numbers express distances to the nodes. The distance Wagner method begins with the two most closely related groups and adds in successively more distantly related groups. Adding in French reveals slightly different amounts of change in English and German:

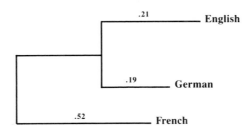

The French lineage cannot be rooted until another language, for example, Grench, is added. Given the following distances:

	English	German	French	Grench
English	0			
German	.40	0		
French	.73	.71	0	
Grench	.91	.89	.90	0

the distance Wagner tree that best fits the data is below.

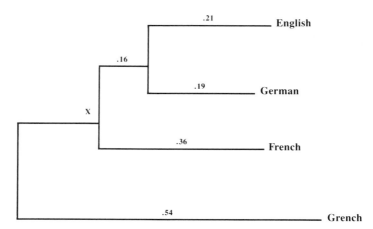

Amounts of change from node X are then fairly equal (.37 to English, .35 to German, and .36 to French). On the other hand, if Grench showed a different pattern of distances:

	English	German	French	Grench
English	0			
German	.40	0		
French	.73	.71	0	
Grench	.80	.78	.99	0

the distance Wagner tree would show unequal rates of change from node X (.30 to English, .28 to German, but .49 to French).

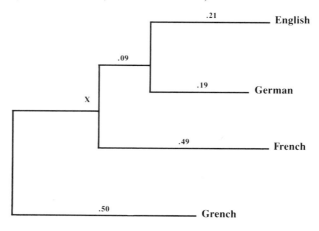

In itself this might be a significant linguistic observation, especially if we could determine whether French underwent an accelerated rate change or whether English and German (or their immediate ancestor) slowed down. In any case, it seems preferable not to assume *a priori* constant rates of linguistic change but to let the data reveal their own structure.

Final Comments

To be fair to historical linguists, it is undoubtedly true that the points made in this chapter are not all unfamiliar. Linguists have been dealing with some of the same sorts of problems biologists have faced for probably the same length of time, incorporating their own special brand of phenomena unique to language change (Hoenigswald 1973; this volume, Chapter 14). It should also be noted that some points discussed herein are not universally recognized by biologists (e.g., the usefulness of not assuming constant rates of change in phenetic clustering method, especially when that is what one wants to prove).

Perhaps the real untapped source of biological-linguistic interchange lies more with outlook than with methodology. In itself, the discussion of the usefulness of cladistics, its underlying philosophy, and its shortcomings by historical linguists and systematic biologists would be a major exchange. Biologists and linguists may also find themselves in comparable situations. For instance, we may agree that after we have drawn our respective phenograms and discovered that rates of change are not constant, instead of being disappointed because divergence times cannot be assigned, one can go on to ask other questions. Can it be determined whether one species or language accelerated or whether its sister group slowed down instead? What kinds of data could discriminate between the two possibilities? Perhaps rough limits can still be placed on divergence events. Do various data subsets alone show the same rate inequalities? If not, is this fact significant for the processes of biological or linguistic change? Maybe shared analytical approaches to the punctuated equilibria model in biology (Eldredge and Gould 1972; Gould and Eldredge 1977) and its linguistic equivalent would be insightful. Since divergence times are so sought-after in both fields, we should consider devising a method for assigning branch lengths to cladograms that would eliminate the distorting component of shared primitiveness. These prospects remain for the future.

Notes

1. The triangle inequality states that for all points a, b, c, and distance function d, where $d(a,b)$ expresses the nonnegative distance between a and b,

$$d(a,b) \leq d(a,c) + d(c,b).$$

2. The time depth formula is given by

$$t = (\ln C) / (2 \ln r)$$

where t is time depth in 10^3 years, C is the proportion of shared cognates, and r is the average proportion of retained basic vocabulary per thousand years, equal approximately to 0.805.

References

Baverstock, P. R.; Cole, S. R.; Richardson, B. J.; and Watts, C. H. S. 1979. "Electrophoresis and Cladistics." *Syst. Zool.* 28:214–220.

Eldredge, N., and Gould, S. J. 1972. "Punctuated Equilibria: An Alternative to Phyletic Gradualism." In *Models in Paleobiology,* ed. T. J. M. Schopf, pp. 82–115. San Francisco: W. H. Freeman.

Farris, J. S. 1970. "Methods for Computing Wagner Trees." *Syst. Zool.* 19:892.

————. 1981. "Distance Data in Phylogenetic Analysis." In *Advances in Cladistics: Proceedings of the First Meeting of the Willi Hennig Society,* ed. V. A. Funk and D. R. Brooks, pp. 3–23. New York: New York Botanical Garden.

————. 1972. "Estimating Phylogenetic Trees from Distance Matrices." *Am. Nat.* 106:645–668.

Gould, S. J., and Eldredge, N. 1977. "Punctuated Equilibria: The Tempo and Mode of Evolution Reconsidered." *Paleobiology* 3:115–151.

Gudschinsky, S. 1964. "The ABCs of Lexicostatistics (Glottochronology)." In *Language in Culture and Society,* ed. D. Hymes, pp. 612–623. New York: Harper & Row.

Hennig, W. 1966. *Phylogenetic Systematics.* Urbana: University of Illinois Press.

Hoenigswald, H. M. 1973. *Studies in Formal Historical Linguistics.* Reidel: Dordrecht.

Hymes, D. (ed.) 1964. *Language in Culture and Society.* New York: Harper & Row.

Nei, M. 1972. "Genetic Distance Between Populations." *Am. Nat.* 106:283–292.

————. 1978. "Estimation of Average Heterozygosity and Genetic Distance from a Small Number of Individuals." *Genetics* 89:583–590.

Ruvolo, M. 1983. *Genetic Evolution in the African Guenon Monkeys.* Ph.D. dissertation, Harvard University. Ann Arbor, Mich.: University Microfilms.

Swadesh, M. 1955. "Towards Greater Accuracy in Lexicostatistic Dating." *Int. J. Am. Linguistics* 21:121–137.

————. 1964. "Diffusional Cumulation and Archaic Residue as Historical Explanations." In *Language in Culture and Society,* ed. D. Hymes, pp. 624–637. New York: Harper & Row.

Wiley, E. O. 1981. *Phylogenetics: The Theory and Practice of Phylogenetic Systematics.* New York: John Wiley.

11

Of Phonetics and Genetics: A Comparison of Classification in Linguistic and Organic Systems

LINDA F. WIENER

Linguistic and organic systems are both characterized by descent with modification through time. There is an unbroken stream of ancestor-descendant relationships with the great majority of characters possessed by one generation passed on unchanged to the next. Lines of descent may split, resulting in different streams of descent for the various parental characters. The task of a systematist or historical linguist is to identify, delineate, and classify taxa in such a way that the subgroupings explain the observed character distributions and yield an understanding of the historical development of the group in question.

It is important to distinguish between a classification and a historical or phylogenetic reconstruction. Classifications can be based on any characters (e.g., all aquatic organisms may be classified together, languages may be classified on the basis of word order), but a historical reconstruction arranges taxa in a manner that reflects genealogical descent. The groups produced through this type of historical classification are natural in that they exist independently of our ability to perceive them. Such groups are discovered, not invented.

A classification may or may not incorporate genealogical information, or it may use both genealogical and other sorts of information (e.g., amount of change since branching) to arrive at a final classification.

There is relatively little open debate about the assumptions, models, or philosophy that underlie linguistic reconstruction techniques, but these questions are topics of intense and lively debate in the systematics community today. Models, assumptions, and philosophies are made explicit, and terms are carefully defined. I wish to compare the procedures used by biologists and linguists in their attempts to construct classifications and understand the historical development of the groups on which they work. These procedures will be discussed in the context of linguistic and organic models of evolution. I am interested in whether the disparities between biological and linguistic techniques stem from real differences in linguistic and organic systems, differences in basic aims or philosophy, or some combination of these factors.

Evolutionary Models

A species may be considered as a group of interbreeding or potentially interbreeding populations separated from other such groups by breeding barriers (Mayr 1969). This is termed the biological species concept. This definition however, is not compatible with all speciation models and all modes of reproduction, for example, it cannot be applied to species with asexual or unisexual reproduction. A newer concept which avoids these difficulties is the evolutionary species concept. An evolutionary species is a single group of ancestor-descendant populations that maintains its identity and is separated from other such groups by virtue of having its own unique history and tendencies (Wiley 1978). Alternatively, ecological niches can be used to delineate species without sexual reproduction (Mayr 1969).

The characters of a species may change. Genetic mutation leads to a change in the phenotype of an organism. This change may be morphological, behavioral, ecological, or biochemical. If the change is advantageous, or at least neutral, it may spread through the population. The new character (apomorphy) coexists at first with the older character (plesiomorphy) and may eventually become fixed. A new character will spread if the individuals possessing it leave more offspring surviving to reproductive age in succeeding generations than those individuals that lack the character leave. Thus, a change in the genotype is reflected in the phenotype of an organism, and the success of a new character is determined by differential survival and reproduction of the individuals that carry the character.

A language may be considered a system of communication separated from other such systems by an intelligibility barrier. Languages may also be formally defined on the basis of possessing different phonological or morphological representations in the underlying language (Agard 1975). However, the connection between a natural language and such a formally defined language is unclear.

Change occurs when an innovation (either borrowed or internal) spreads through the speakers of a language. Phonological shift is generally considered to be the prototypical language change. There are two aspects to the spread of a phonological innovation. A change must spread through all speakers and all words with the phonetic environment in question. Sound change is supposed to be exceptionless, that is, it affects every word with the phonetic environment in question regardless of function. Grammatical, semantic, and syntactic changes are often driven by a preceding sound change (Bynon 1977). These changes may reach all speakers, but they do not necessarily affect all possible environments.

Language change starts externally as a variant of an accepted form and moves through populations of speakers who choose whether or not to use it and under what circumstances. Children learning the language construct their own grammars on the basis of the language they hear around them. An innovation that is purely optional for the parent generation may be interpreted as the norm and incorporated into the child's grammar as a rule. Thus, the rules of a child's grammar may differ considerably from those of the adult generation and may be passed on to subsequent generations.

Languages do not adapt in the way in which organisms can be said to adapt. The success of a language is not dependent on the characters of the language itself, but on the social status of the people who speak it. Thus, new characters spread through a language if they are perceived as being associated with a prestige group. Also, the language characters of the parents do not usually determine the first language of their children. The peer group of a child is largely responsible for determining the characters of a child's speech. Thus, if parents are native speakers of German and move to America their children will be native speakers of English.

Methodologies

There are currently three active schools of organic taxonomy: phenetics (Sneath and Sokal 1973), in which classifications are based on measures of overall similarity; cladistics (Hennig 1966), in which classifications are based on possession of shared derived (synapomorphic) characters as distinguished from shared primitive (symplesiomorphic) characters; and evolutionary taxonomy (Mayr 1974, Simpson 1961), in which a number of different types of information are considered involving both phylogenetic and other types of characters to arrive at a final classification.

Historical linguists have generally used some variant of the comparative method since publication of the neogrammarian manifesto (Osthoff and Brugmann 1878). The comparative method is used to establish classifications based on possession of shared innovations and so is similar to cladistics in philosophy. Swadesh (1951) developed a method termed glottochronology in which branch points are established based on the number of cognates among certain specified basic vocabulary items. This method does not distinguish between synapomorphies and symplesiomorphies and so is similar to phenetics in philosophy.

I will be concerned with cladistic techniques in biology and the comparative method in linguistics for the remainder of this chapter. Both methods involve the construction of trees with branch points based on possession of synapomorphic characters. Similarities due to chance resemblances and parallel developments (homoplasies) are screened out. When there is a conflict between characters, the most parsimonious tree (e.g., the tree that assumes the fewest number of steps and the fewest ad hoc hypotheses) is preferred.

With this basic outlook in mind, I will illustrate some of the similarities and differences of organic and linguistic systems by comparing problems associated with the identification of primitive versus derived characters. I will consider monophyly, identification of homologues, use of the historical record, and hybrid taxa.

Life is usually considered to be holophyletic, that is, it arose once and all species are descendant from the first one. Successive monophyletic branches can be established on the basis of synapomorphies until all taxa are resolved. Each new branch has characters that are relatively more derived than the one it split off from (its sister group). These characters can be used to reconstruct phylogeny. Monophyly is an important assumption because it is the basis of the outgroup method. This method involves the assumption that a sister group largely retains the primitive condition. Decisions about the polarity of character states can be made by using the sister group to estimate what the primitive character was like. Representatives of ancient taxa have often survived to the present; for example, algae are living primitive plants, cockroaches are living primitive insects, and so on. This greatly aids in the reconstruction of phylogeny above the species level. However, accurate identification of the sister taxon, especially at higher levels of classification, may be difficult. Decisions about the sister group may affect the classification in important ways. This problem may be circumvented by rooting the tree after the characters are resolved (Farris 1972).

There is no good evidence for or against the holophyly of language. A single origin cannot be assumed with confidence because the relevant characters have been lost with the passage of time. Relationships can be analyzed only when a common ancestor can be confidently posited on the basis of recurring correspondences between sound and meaning in basic vocabulary items and grammatic paradigms. Representatives of primitive taxa do not survive to the present, so an outgroup method is not practicable for analyzing phylogeny. Linguists estimate the primitive character by reconstructing features of the common ancestor (protolanguage) of the group in question. This is done by using evidence from extant languages and/or records from dead languages to hypothesize what the sounds, semantics, grammar, syntax, and lexicon of the ancestor were like.

Reconstructing protoforms from the evidence of daughter languages and then constructing subgroupings on the basis of these protoforms may involve aspects of circularity. Much sound change is hypothesized to be unidirectional, but it is unclear how constrained sound change actually is. However, the comparative method has proved to be a strong predictive tool. It has been vindicated on a number of occasions when a very archaic member of a language family has been deciphered. Notably, Mycenean Greek (Linear B) from the Indo-European family contained many of the features predicted for Proto-Greek. This predictive power even extends to characters that were hypothesized to exist in the protolanguage even though none of the daughter languages examined contained the character in question (e.g., the laryngeals of Hittite).

A further objection to using the procedure of reconstruction to estimate primitive characters is that it is typological (in the biological sense in which a type or standard was thought to exist for any given species, with aberrant individuals seen as variants of the type). This notion is demonstrably inconsistent with the real world, in which great variability in both species and languages is always observed.

The problem of variation is most serious at the species and language levels. There is a range of variation for any given character. At higher levels of classification, however, the characters that are informative cladistically can be considered fixed and invariant. The outgroup is used to estimate these informative characters in biology. Linguists reconstruct the stem group itself, the actual ancestral taxon. This ancestor was obviously not invariant. Variation is the raw material of change itself and must have been present in order for the various daughter languages to develop.

In a given language area there may be a prestige dialect that is perceived as a standard. People can and do strive to speak that dialect, and there may be considerable reward in the form of social standing and higher paying jobs for doing so. Thus "typology" (in the sense of a standard) is a more realistic concept in linguistics than in biology, in which aberrant individuals cannot strive to be like an ideal type. This point is especially relevant when written records are considered. The written language is much more formal, much more standard than the spoken language. This standard form is the only record we have of most dead languages.

A key problem in both fields is the ability to identify homologous characters. In biology, similarity of position, appearance, function, and ontogeny are used to make hypotheses about homology. Various factors act to obscure proper identification of homologous characters. Adaptation is the major factor. Similar characters evolve as solutions of similar problems of survival (e.g., a spherical shape and spiny leaves have evolved in distantly related plant groups to solve the problem of survival in desert environments). Such parallel adaptations are even more likely in closely related taxa that have similar genetic starting materials. Characters may converge simply by chance as well, with no obvious adaptive explanation. Hypotheses of homology are tested by constructing cladograms. Similar characters that are consistent with the most parsimonious cladogram are accepted as homologous.

Cladists do not treat all characters as equally informative. Characters that are not subject to strong selective pressures and that are relatively complex and therefore unlikely to evolve twice simply by chance are given more weight in the analysis (e.g., wing venation and genitalia structure in insects). Decisions about relative weight are usually made on an *a posteriori* basis after considerable experience with the group in question. It is important to note that any given character is informative at only one level of the branching diagram, and a character that seems uninformative may simply have been considered at the wrong level of universality.

Linguists use aspects of phonology, grammar, semantics, and syntax to make hypotheses about homology. Generally, a phonological analysis is necessary before grammatical characters can even be correctly identified. Thus, phonological features are often considered the primary evidence for identification of synapomorphies. However, sound change often produces forms that are difficult to pronounce or that result in irregular grammatical paradigms. These new features are frequently leveled out in order to simplify or regularize the language once again. This type of functional reanalysis is termed analogical leveling and is the major process that acts to obscure homology in language groups. It is not *a priori* predictable if an irregularity or awkward combination of sounds will be leveled, and if so, how. Irregularities in core vocabulary often survive for very extended periods of time and are useful in identifying remote relationships.

Borrowed features can also confuse analysis of homology. It may not be clear if a character is inherited or borrowed from a closely related language. Accepting a borrowed feature as native may result in hypothesizing a nonexistent sound change or grammatical form. A borrowed character may be screened out by knowledge of the source language or by tracing a form back to its corresponding form in the protolanguage. If the character is inconsistent with those of the protolanguage, it is likely to have been borrowed at some point subsequent to the split of the protolanguage.

Identification of homologues, and proper designation of primitive versus derived character states, is usually considerably easier at the level of the species or language than at higher levels of classification. A great deal of change accumulates in branches that diverged long ago, and it becomes very difficult confidently to identify homologous characters of any value. Hybridization and mosaic evolution further complicate this matter. A hybrid taxon that spawned a lineage in remote times may be impossible to identify as such. In language this is especially difficult because two characters may be homologous but one (or both) may be borrowed rather than inherited from a common ancestor.

Both linguists and biologists can get clues from historical records of taxa that are no longer living. The record for both languages and organic groups ranges from nonexistent to quite complete. Biologists and linguists differ in their treatment of evidence from the historical record.

Cladists do not give special status to fossils. They recognize that much less information may be available about fossils than about extant organisms. Older fossils do not necessarily have more primitive characters; therefore, fossils do not have a privileged position in establishing phylogenies. Their characters are analyzed in the same manner as those of living taxa.

Written records are given special status in linguistic analysis. Generally, the oldest available records are weighted most heavily in the reconstruction of protolanguages. This is because newer representatives of a linguistic group have lost many of the archaic characters that were contained in the proto-language. It is recognized that dead languages are often "derived" for certain features and that they may be contaminated by borrowings from other languages, but consideration of data from all branches is generally helpful in spotting such characters. Linguists may have sources of information other than the scripts of a dead language to work with. Further information can be gleaned from monolingual, bilingual, and multilingual glossaries and texts as well as translations, transliterations, and comments left by ancient grammarians and explorers. These sources must be evaluated critically, but they can be enormously useful.

At this point I turn to the special problem of hybrid taxa and the difficulties associated with classifying them. The models of linguistic and organic evolution diverge most noticeably when hybrids are considered. The formation of hybrids in organisms indicates that two supposed species were not really separate, or at least that a close relationship exists between the hybridizing species. The characters of these hybrids generally correspond in form and function to the characters of the parent species and are often intermediate between those of the parent species.

All languages are hybrid to some extent. A close genealogical relationship is not required for hybrid formation. Lexical items move especially easily from one language to another, but phonetic, grammatical, and syntactic features may also be borrowed.

Several types of hybrids are possible. All languages contain many borrowed features and are therefore partially hybrid. This type of hybridization is best termed introgression and distinguished from other more completely hybrid taxa.

Bilingual and multilingual individuals carry all the characters of the languages that they know. Though the languages are usually kept separate in the speaker's minds, these individuals can move characters from one language to another. This is important in the appearance of areal features that are traits shared by languages in a geographic area, regardless of genealogical relationship (e.g., the clicks of certain African languages).

Pidgins and creoles are languages that blend the characters of two or more parent languages to form a new language. The features of these languages are not generally the same in form or function, or intermediate between those of the parent languages. Rather, they are a conglomerate of features selected from each parent and often used in unique ways. I will not discuss the complexities of pidgin and creole classification, but will confine myself to the most common case in which a language has many borrowed features.

A consequence of the large number of adopted features that languages acquire is that a tree-type classification is not maximally predictive of shared information, as is claimed for a tree generated by cladistic analysis of a group of organisms (Farris 1980). A tree analysis of a language group reflects only those characters that can be traced back to a common ancestral language. Although English has many French features and can be construed as descending from both Romance and Germanic ancestors, English is always classified with the Germanic languages because it has synapomorphies with German, Swedish, and so on, but has no synapomorphies with French and was never part of a speech community that later split into French and English branches.

Borrowed characters present special problems in the identification of primitive and derived conditions. A language can innovate both by creating new features or by borrowing them. Borrowed features must be screened out if they were acquired after the period for which subgroupings are being established. Characters that were borrowed by a language before it split can be analyzed in the usual manner.

Linguistic evolution is only partly hierarchical. The comparative method deals with this aspect of language change. There have been no good, realistic methods for dealing with both the hierarchical and the reticulate aspects of language evolution, although both are important aspects of language change.

Cladistic analysis assumes a hierarchical pattern. This may be quite unrealistic for certain groups (e.g., bacteria). It is important to realize this, and not to force a tree-type classification on such groups simply for convenience. It may be possible, as in language analysis, to separate and appropriately treat these different aspects of evolution.

When all the foregoing comparisons are taken into account, it becomes apparent that certain problems are indeed common to both disciplines, and certain are clearly special to each. Some areas of comparison that should prove especially fruitful are variation, identification of homologues, and methods for dealing with both the hierarchic and reticulate aspects of character evolution. A comparison of various proposed solutions may be useful in clarifying and resolving these difficulties.

It is also important to note the very real differences in organic and linguistic systems. Some of the most important differences are that primitive languages have not survived to the present day, while many primitive organisms have. Languages are all hybrid to some extent, while most organisms only occasionally produce hybrids. Also, there is no necessity for close genealogical relationship for hybrid formation in language groups, while this is essential in organisms. The features of a language are not adaptive and do not determine its relative success. Rather, the spread of a language character is dependent upon the social status of the people who use it. This is quite different from a species, in which inherited characters may directly affect an individual's relative success.

One aspect of this comparison that has not been adequately dealt with is the difference in basic approach of systematists and historical linguists. The actual techniques used to infer phylogeny in both fields are very sophisticated, but the theoretical framework in which these techniques fit is much more explicit and therefore much more open to debate in biology than in linguistics. In particular, many central concepts such as the definition of language and how to determine when two languages are the same (both synchronically and diachronically) are little debated in the context of how they affect various theories of language evolution and classification. Greater attention to explicit definition of terms and explication of models would greatly facilitate understanding.

In answer to my initial inquiry, the disparities between biological and linguistic techniques stem from real differences in linguistic and organic systems as well as differences in basic aims and philosophy. Both the similarities and differences of these systems must be kept in mind if useful comparisons are to be made. However, there is much room for fruitful comparative work that may yield new or improved techniques and a better understanding of evolutionary processes in both linguistic and organic systems.

Acknowledgments

The author would like to thank all those who attended the conference for their discussion and comments. Special thanks go to Peter Stevens and Ernst Mayr for reading and commenting on various versions of this chapter.

References

Agard, F. B. 1975. "Toward a Taxonomy of Language Split (Part One: Phonology)." *Leuv. Bijd.* 64:293–312.

Bynon, T. 1977. *Historical Linguistics.* Cambridge: Cambridge University Press.

Farris, J. S. 1972. "Estimating Phylogenetic Trees from Distance Matrices." *Am. Nat.* 106:645–68.

———. 1980. "The Information Content of the Phylogenetic System." *Syst. Zool.* 19:83–92.

Hennig, W. 1966. *Phylogenetic Systematics.* Urbana: University of Illinois Press.

Mayr, E. 1969. *Principles of Systematic Zoology.* New York: McGraw-Hill.

———. 1974. "Cladistic Analysis or Cladistic Classification." *Z. Zool. Syst. Evolut.-Forsch.* 12:94–128.

Osthoff, H., and Brugmann, K. 1878. *Morphologische Untersuchungen* I. Leipzig: Hirzel.

Simpson, G. G. 1961. *Principles of Animal Taxonomy.* New York: Columbia University Press.

Sneath, P. H. A., and Sokal, R. R. 1973. *Numerical Taxonomy.* San Francisco: W. H. Freeman.

Swadesh, M. 1951. "Diffusional Cumulation and Archaic Residue as Historical Explanation." *S. W. J. Anthrop.* 7:1–21.

Wiley, E. O. 1978. *Phylogenetics.* New York: John Wiley.

12

The Upside-down Cladogram: Problems in Manuscript Affiliation

H. DON CAMERON

In 1977, Norman Platnick of the American Museum of Natural History and I published an article in *Systematic Zoology* drawing attention to the similarities of method in historical linguistics, textual criticism, and phylogenetic systematics. We pointed out that methods of constructing and testing hypotheses about the interrelationship of taxa connected by ancestor-descendant relationships were developed by classical scholars systematically in the early nineteenth century and that the origins of the method can be found as early as the sixteenth century in the manuscript studies of Erasmus of Rotterdam and Joseph Justus Scaliger.[1] All three fields use analogous procedures in which data are organized into transformation series of homologous character states, the polarity of these transformation series is determined by out-group comparison, and shared innovations are used to construct internested series of three-taxon statements.

The texts of ancient authors survive through successive copying over a span of 2,500 years. Errors are bound to creep in, so that the extant manuscripts of an ancient author, say the plays of Aeschylus or Euripides, will differ significantly from one another. To make this clear, let me sketch out an idealized story of how an ancient text gets transmitted to us.

In the last third of the fifth century B.C., Euripides produces a finished copy of the *Medea* and uses this copy to train the actors and choristers, who present it in the spring dramatic contests at Athens in 431 B.C. As a result of its success, copies are made by booksellers, and these copies can be bought in Athens, exported, and widely circulated.[2] In the next century, repertory companies revive the play, but adjust it to their needs by interpolating and omitting lines (Page 1934). The variation in the circulated texts is such that, in 330

B.C., Lycurgus carries a public decree that an official copy of the works of Aeschylus, Sophocles, and Euripides be made and kept in the public archives and that the actors should correct their texts against it.[3]

When Ptolemy II Philadelphus established the great library at Alexandria around 280 B.C., he acquired by a certain amount of skullduggery this official Athenian state copy.[4] From it the great Alexandrian scholars produced their learned editions and commentaries. Once again the book trade disseminates these texts. In the Roman period an edition of ten selected plays, which includes the *Medea,* is made. Up to this time the text has been on rolls of papyrus, but by the fourth century A.D. it will have been transferred to the parchment codex, with the result that it is possible to lose leaves and quires, and the order of the plays will make a greater difference (Roberts 1954; Widmann 1967, p. 587ff). In the ninth century there was another burst of scholarly activity, the imperial university was revived at Constantinople (Ostrogorsky 1969, p. 224), and a new interest in the classics flowered. Two technical developments occur: paper becomes common and the texts are transliterated from majuscule script (i.e., what we would call capital letters) into minuscule script, a cursive and space-saving book hand.[5] A second revival of learning in the fourteenth century provides us with many of our most important manuscripts. For Euripides and Aeschylus in particular, a name to be honored is Demetrius Triclinius, without whom we would lack the *Agamemnon* and nine of the plays of Euripides. In the fourteenth century, Greek manuscripts began to find their way to Italy. The great Laurentian manuscript of Euripides was brought to Avignon by Petrarch's teacher of Greek, Barlaam, sometime before 1348 (Zuntz 1965, pp. 281–288). By the early part of the fifteenth century, there was a concerted effort to bring Greek manuscripts to Europe. For example, in 1423 Giovanni Aurispa brought a collection of 238 Greek books to Europe (Sabbadini 1905, 1:47), and it is from such manuscripts that the early printed editions are made beginning at the very end of the fifteenth century.

Consequently, the physical evidence for the texts of Greek antiquity consists of medieval codices of the ninth through the fifteenth centuries, found now largely in European libraries, and it is from these manuscripts that we reconstruct (ideally) the very words, the *ipsissima verba,* that Euripides wrote in his study two thousand years previously. With all the possibilities for error consequent upon the span of time and the jarring disturbances in the course of the tradition, it is a wonder that we succeed. But it turns out that we can be reasonably confident that we come fairly close. For instance, quotations from Euripides in other ancient authors (e.g., Plato, Aristophanes, Aristotle) tend to confirm the text derived from the medieval tradition, and so do papyri dated as early as the third century B.C., such as the famous Strasbourg papyrus containing fragments of the *Medea* (Page 1938, p. xlix).

The theory of constructing tree diagrams representing the relationships of manuscripts is in all essential points the same as the cladistic methods of phylogenetic systematics, and the classic exposition of the theory is that of Paul Maas, first published in German under the title *Textkritik* in 1927 and since translated into English with the title *Textual Criticism* in 1958. Maas might be called the Hennig of manuscript studies. As one might suspect, there is also a school that vigorously demurs against the theory. We call the method "stemmatics" and the resulting diagram a "stemma."

In the course of successive copying, errors are bound to occur, and once an error lodges in a manuscript that serves as an exemplar from which one or more descendant manuscripts are copied, the error will occur also in the derived manuscripts. It will endure—at least it will endure if it is a nice, well-behaved error of the kind we like to see. We will not be interested in the kind of error that happens all the time, everywhere, spontaneously, such as the confusion between *epsilon iota* and *eta,* or between *omicron* and *omega,* for such adventitious errors tell us nothing about the relationships of manuscripts. Such errors present problems analogous to convergence and parallelism in evolution. We will also not be interested in errors that are so obvious that it is easy for the copying scribe to correct them, so that the absence of the error does not prove the manuscript was not copied from an exemplar that exhibited the error. The analogue here is retrograde evolution, the ultimate disappearance of a new apomorphous character, such as the loss of wings in fleas.

Well-behaved errors, those that can be used to make inferences about manuscript relationships, are called "indicative errors" (Maas 1958, p. 42). The judgment whether a given error is likely to occur spontaneously in two unrelated manuscripts or is likely to be easily corrected by the scribe, and the judgment whether an error is likely to indicate the dependence or independence of manuscripts, amounts to a weighting of the features.

As in cladistics, shared innovations are used to prove that manuscripts descend from a common exemplar; that is, if two manuscripts A and B share an indicative error against a third C, it proves first that C is not dependent upon either A or B, and that A and B belong together on the same branch of the stemma. Such an indicative error is called "conjunctive," and implies the following three possibilities:

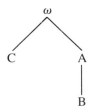 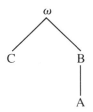 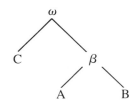

We can sometimes eliminate one of the possibilities by extrinsic historical evidence. Or the material evidence of the manuscript may serve to prove that one of the manuscripts was copied from the other. Or the objective dating of the manuscripts may eliminate one of the possibilities.

But we may resolve the question logically by finding two *separative* errors, that is, a pair of errors that prove that neither A nor B could have been copied from the other. If A has an error not found in B, then B was not copied from A. Contrariwise, if B has an error not found in A, then A was not copied from B. The pair of separative errors demonstrates that the third stemma is correct. Greek letters are conventionally used to designate lost hypothetical ancestor manuscripts, and thus β is the hypothetical ancestor of both A and B. This means that we can reconstruct the readings of β, specifically that we can claim that the conjunctive error found in A and B was read also in the hypothetical β. Further, the same conjunctive error can serve *again* as a separative error proving that C was not copied from the hypothetical β. We would then look for a symmetrical separative error in C which would prove that β was not descended from it.

It will be readily apparent that a conjunctive error is a synapomorphy, and that a separative error is an autapomorphy. No matter how many manuscripts are to be included in the tree, the analysis may be dissected into a series of three-witness problems. For example with a fourth manuscript we may get a tree like this:

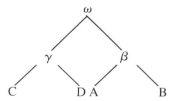

The extant manuscripts are placed at the end-points of the tree, and the hypothetical manuscripts, designated by Greek letters, are found at the nodes. There is no logical reason why a real manuscript might not be placed at a node, and in fact we can sometimes do that. A famous example is Hermann Sauppe's proof in 1841 that all extant manuscripts of the speeches of the Athenian orator Lysias are derived from the twelfth-century Codex Palatinus (Sauppe 1841). When a real manuscript appears at a node, we may remove from consideration all the manuscripts dependent upon it, since they contain no information that is not already contained in their ancestor.

We cannot simply assume without proof that the manuscripts of a given author actually do descend from a single archetype,—ω in our stemma—but this must be explicitly demonstrated.[6] To do so we must find an error that all extant manuscripts share. Some early triumphs of the method were precisely directed toward this end. In 1508 Erasmus argued that an error in all the manuscripts was proof that they descended from a common archetype,[7] and Joseph Scaliger argued in 1552 from the nature of the errors in the manuscripts of Catullus that they derived from a common parent.[8] Theoretically this is a step forward from Politian's argument that a particular *extant* manuscript is the ancestor of the others, to the postulation of an archetype that no longer exists.

But I have been talking very casually of errors. How do we know we have an error? Outgroup comparison will help, once we have established the stemma, but we need to determine at least some of the indicative errors before we can ever do that. Errors are equivalent to apomorphies, and correct readings are equivalent to plesiomorphies, but given two readings, how do we determine the polarity of the feature? It may be noted here that with texts the problem of establishing homologies, which is a major difficulty for the zoologist, is trivial, since a written text consists of a strictly ordered linear sequence of elements.

The plainest kind of derived character results not from an error in copying but from some physical damage to the exemplar, such as loss of pages. Sauppe's proof that all manuscripts of Lysias depend upon Codex Palatinus 88 (Sauppe 1841) rests on the fact that the table of contents in the front of the codex lists a speech that is not there. Furthermore, the end of the preceding speech and the beginning of the following speech are missing. This loss corresponds exactly to a quaternion that has clearly fallen out. The dependent manuscripts all omit the same stretch, but without the coincidence of page ending and text and without the concomitant physical loss.

But we are rarely given the luxury of such obvious physical causes, and the vast majority of errors result from faulty copying, the product of the habits and psychology of the scribes. The determination of errors ultimately rests on the high degree of redundancy in a text, that is, any part of the text is supported by grammatical and logical coherence, orthographical convention, or metrical law to a degree that makes the detection of errors possible.

For example, in Euripides' *Trojan Women* a choral passage of regular iambic dimeter (551–554) goes:

$$\breve{\;}\;\bar{\;}\;\breve{\;}\;\bar{\;}\;\breve{\;}\;\bar{\;}\;\breve{\;}\;\bar{\;}$$
ἐγὼ δὲ τὰν|ὀρεστέραν

$$\breve{\;}\;\bar{\;}\;\breve{\;}\;\breve{\;}\;\breve{\;}\;\breve{\;}\;\bar{\;}\;\breve{\;}\;\bar{\;}$$
τότ᾽ ἀμφὶ μέλα|θρα παρθένον

$$\breve{\;}\bar{\;}\;\breve{\;}\;\breve{\;}\;\breve{\;}\;\breve{\;}\;\bar{\;}\;\breve{\;}\;\bar{\;}$$
Διὸς κόραν|[Ἄρτεμιν]|ἐμελπόμαν

The meter would be fine without the word Ἄρτεμιν, which I have bracketed.

The passage means: "And then throughout the halls I sing the praises of the mountain maid, the daughter of Zeus, Artemis." It is clear that "Artemis" is an explanatory note, once written above the line, or in the margin, to explain just which daughter of Zeus is meant. That is, it is a lexical gloss written by a scholiast to aid the reader, which was then at the next stage of copying understood to be an omission that ought to be restored into the text. The combination of the metrical irregularity, and the rationale that the word is a gloss, makes it virtually certain that the word should be omitted from the text.

The only manuscript of the *Choephoroi* of Aeschylus reads at line 936 καρύδικος ποινά, "punishment which is nut-just." Obviously "nut-just" means nothing, and is likely the result of a copying accident. It is a familiar kind of accident, because minuscule manuscripts constantly confuse *beta* and *kappa,* since their cursive forms are very similar. If we substitute a *beta* we get the word *barydikos,* which makes perfect sense, meaning "bringing heavy justice."

From these two examples it is possible to see that there are two steps in detecting an error. The first is to notice some illogicality in the text, such as a metrical irregularity, spelling error, grammatical inconsistency, meaningless words, lack of grammatical or logical connection, or anachronistic grammatical forms or diction. From these more or less objective criteria we proceed to errors of style judged by comparison with the rest of the author's work, and fairly subjective matters of taste and decorum. The second step in detecting errors is to explain how the error came about, that is, to make a hypothesis to argue that such an error could actually have taken place and is likely to have taken place in the present instance. To do this we have developed a detailed typology of errors, by which any instance can be placed into a wider theory of error. Errors range from simple confusion of letters to false expansion of abbreviations, false word division, substitution of synonyms, substitution of whole words of similar appearance, interpolation of glosses, leaping from one instance of a word to another leaving out the intervening text, omission of line through similar beginnings or endings, transposition of words and passages, mistakes due to changes in pronunciation, haplography, dittography, and insertions from the margin. That is to say, if we can rationalize the ontogeny of an error, we have evidence for the polarity of the feature.

There is a famous example of how the ontogeny of a single error was sufficient to determine the relationship of two manuscripts. For generations there has been an argument about two of the most important manuscripts of Euripides, L and P. Which is dependent on the other? Both are fourteenth-century manuscripts; both are from the same workshop. L is paper; P is vellum. On June 3, 1960, Gunther Zuntz was examining L in the Laurentian library in Florence at *Helen* 95. In the other manuscript P the line is written like this:

πῶς· οὔτι που σῷ φασγάνῳ βίον: στερεῖς

The colon near the end of the line is placed inexplicably in the middle of a phrase. The corresponding colon in L seemed to be of a peculiar color, and Zuntz examined it under an ultraviolet lamp. He asked the librarian, Dr. Anna Lenzuni, for her opinion. "She ran her hand over the place—and the "colon" stuck to her finger. The heat of the lamp had loosened it. It was a tiny piece of straw . . . imbedded in the coarse paper." The scribe of P had punctiliously copied this bit of straw as a colon, and this proves absolutely that P was copied from L. The piece of straw is kept in a tiny box in the safe at the Laurentian Library as the decisive piece of evidence (Zuntz 1965, pp. 14–15).

There are times when the tradition offers two readings that are both acceptable, that is, neither is logically or grammatically objectionable. It may be possible to determine which is apomorphic on the basis of outgroup comparison. The manuscripts of the *Medea* give us sufficient evidence to construct a secure stemma, and we know from an error in line 1359 shared by all of them that they descend from a common archetype and represent a closed tradition. The stemma looks like this:[9]

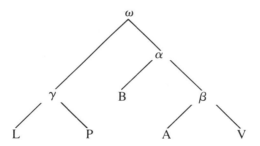

At line 912 LPB read ἀλλὰ τῷ χρόνῳ and AV read ἀλλὰ νῦν χρόνῳ. Both mean something like "but at last." The reading of LPB can be better supported by parallel passages, but it is possible to make an argument for the reading of AV. But outgroup comparison decides the matter, and we must regard the agreement of LPB as proof of the reading of the archetype. Add the fact that we can explain the reading of AV as a gloss, introduced in β, and no doubt remains.

Manuscript errors sometimes form what zoologists call a transformation series, that is, at a given point of the text the manuscripts may exhibit errors that form a directed series or succession of character states. To put it another way, the available manuscript readings of a given passage may be arranged in a series with the true reading at one end and a progressive succession of errors with the most deviant at the other. Suppose we have a succession of readings, beginning with the true reading, A, B, C, D, such that the conditions for error C do not exist in the true reading A but are generated by the error in B, and further down the line the conditions for error D did not exist until error C was made. Each error results in new conditions that are necessary for the succeeding error, and if any intermediate stage were missing, say B or C, the chain would be broken, and we would not be able to explain the error D. If we can discover such a transformation series, we have strong evidence for the relationship of the manuscripts in which the errors occur. Of course, such a chain does not always prove that D is dependent on B, but only that D is dependent on a manuscript, perhaps hypothetical, from which B descends.

The Hypothesis to the *Medea* provides an example to make this clear. The medieval manuscripts of ancient drama will often have material introductory to the text of the play, summarizing its plot and giving information about its date, the conditions of its original performance, or some gossip about its origins. These Hypotheses originated with the Alexandrian commentators, like Aristophanes of Byzantium, and their successors (Zuntz 1955, 129–152). A passage in manuscripts A P D F of the *Medea* tells us that Euripides seems to have adapted his play from an earlier play of Neophron. The true text reads τὸ δρᾶμα δοκεῖ ὑποβαλέσθαι παρὰ Νεόφρονος and so it stands in D. The prepositional phrase παρὰ Νεόφρονος ("from Neophron") undergoes a series of changes in the other manuscripts as follows:

F   περὶ Νεόφρονος   "about Neophron"
P   παναιόφρονος   meaningless
A   γενναιοφρόνως   "in a noble minded fashion"

These errors indicate the following stemma:

In ω the preposition παρὰ was written out fully, but in ε it was abbreviated with a common contraction πᵉ. This was misunderstood in two ways: F misread it as *peri* and π copied it as πα. A second error in π misspells Νεόφρονος as Ναιόφρονος by the identity in sound. P copies faithfully the meaningless παναιόφρονος of π, but A tries to amend it into something meaningful. Without the preposition, there is no longer any need for a genitive, so A amends it into an adverb, and fixes up the front of the word to produce the plausible word γενναιοφρόνως. Without the intermediate stages, and without the true reading in D, supposing we had only the final stage in A, it would be impossible to work back to the true reading.

Let us now turn to a sketch of the historical development of these methods to help to keep in view the purposes which the methods are meant to serve.

When the first printed editions of Greek texts were made beginning with the end of the fifteenth century, for example, by Aldus Manutius, the procedure was to find a manuscript, any manuscript, and print it. For instance, the Aldine Aeschylus (1518) was printed from a fifteenth-century manuscript (Guelpherbytanus 4725) that is now properly disregarded by editors, since it is merely a copy of the tenth-century Mediceus, which exists. Accident and chance often determined what manuscript was used for early printed editions.

But when manuscripts of a text were numerous, and standards of editing could be improved, editors were presented with the practical problem that the number of manuscripts was simply unmanageable and some principle of selection was necessary. Often editors chose what they considered to be the oldest manuscript and edited it freely with many conjectural emendations. Extreme confidence in the powers of conjecture was memorably voiced by Richard Bentley (1662–1742) in his notorious dictum "Nobis et ratio et res ipsa centum codicibus potiores sunt" (For us reason and the context are stronger than a hundred manuscripts).[10]

In reaction to this, Friedrich August Wolf, regarded as the father of modern classical philology, warned that, where all manuscripts agree, the editor should not depart from them except for the most compelling reasons (Wolf 1795; quoted in Hall 1913, p. 122). He says that all trustworthy witnesses to a text must be heard, and heard continuously before decisions are made. Wolf's pupil Immanuel Bekker (1785–1871) collated about four hundred manuscripts in his lifetime, but tended to regard the oldest manuscript as the best, or regarded the best family of manuscripts as the only authority to be considered. Timpanaro (1963, p. 74) compares this early reliance on the best or oldest manuscripts to the conviction of some early Indo-European linguists that Sanskrit always preserves the more ancient state.

But at the same time, following the example of their Renaissance forebears Politian, Erasmus, and Scaliger, others aimed at reconstructing an archetype from logical analysis of the existing manuscripts. Bentley had proposed an edition of the New Testament on such principles, and Karl Lachmann likewise made a tentative recension of the *Nibelungenlied* on stemmatic principles as early as 1817. But it was the Danish scholar Johan Nicolai Madvig who in 1833 produced the first full-scale cladistic argument on the relationship of manuscripts of Cicero's *Pro Sestio* and *In Vatinium*.

Such analysis enables the editor to eliminate from consideration all manuscripts that do not carry independent information about the archetype. The principle of selection is not the age or even the quality of the manuscript, but the independence of the information it carries. Once we can reconstruct the reading of the archetype we have an important check on the extravagancies of conjectural emendation.

The possibility of making a stemma of extant manuscripts depends on the assumption that each manuscript is copied from precisely one exemplar, rather than conflating two traditions from two or more exemplars. Sometimes a manuscript appears to be "contaminated." The stemma of the manuscripts of Herodotus will illustrate the point.

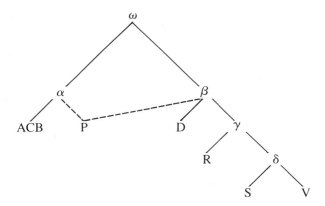

This is as pretty and as well behaved a stemma as one could wish, and the manuscripts fall into line like dutiful soldiers. We can reconstruct with confidence the two major hypothetical variant carriers α and β. But there is a complication. Manuscript P is not well behaved. We can find conjunctive errors between P and A C B, but also between P and D R S V.[11] P is a hybrid manuscript and presents the same kinds of problems as hybrid species in biology.[12] This contamination appears to result from scribes copying eclectically from two manuscripts that derive from different branches of the stemma.

There is an increasing amount of evidence of what we call horizontal transmission, where characteristic errors of several branches turn up in a single manuscript. This happens not only when scribes compared different copies, and selected readings from them, but also when variant readings that may be cited in the margin or in scholiastic commentaries find their way into the text, either by scribal choice or by misreading a marginal variant as a correction. The manuscript tradition of Xenophon's *Cyropaedia* has been so contaminated that the manuscript relations are positively reticulated.[13] The investigations of R. D. Dawe (1964, 1973) on the texts of Aeschylus and Sophocles argue that stemmatic method for these authors is virtually worthless. He argues that the simple task of excising the hybrids is hopeless. In his book on Sophocles he says, "We have seen that the relationships of the MSS in question cannot meaningfully be represented in stemmatic form, even if we make the most generous allowances for contamination obscuring the basic affiliations. We have seen that the stemma is not only wrong, but wrong in every detail" (Dawe 1973, 2:34). Dawe expends enormous labor to show that true readings constantly show up in dependent manuscripts of Sophocles and Aeschylus, readings that are not found in any other manuscript. That is, an inferior manuscript may preserve the true reading. Dawe assumes that this can happen only if the true reading has been imported horizontally into the inferior manuscript (rather than by lucky conjecture, for instance), and since this happens often, it must be the case that our theory of transmission has been wrong. "We believe that the fact of unique preservation has been demonstrated; consequently the fault must lie with the theory of descent, and we conclude that the . . . stemma does not after all represent, even in the simplest form, the true character of the tradition" (Dawe 1964, p. 157).[14]

Dawe denies radically that archetypes can be reconstructed, but he necessarily pays a theoretical price for his conclusion. In order to argue that the tradition of Aeschylus' plays is "open," that is, not descended from a single archetype, he must discredit the importance of indicative errors shared by all the extant manuscripts. For example, there are three instances of line displacement in the *Seven Against Thebes* which have been regarded as reflections of the single archetype. Dawe says, "The three places cited from the *Septem* no doubt do each preserve a memory of the arrangement in a particular ancient manuscript, but it would be irresponsible to assume that in all three cases it was the same ancient manuscript that was involved." And he further says, "We must be on our guard against one of the elementary errors which stemma drawing has tended to exalt into a prime virtue, namely the confusion of economy with truth" (Dawe 1964, p. 159). That is, in order to discredit the concept of an archetype, he must explicitly reject any criterion of parsimony.

If there are no archetypes or stemmata, and if true readings are uniquely preserved in any manuscript regardless of its stemmatic position, we are then thrown back to a procedure of evaluating readings which is unaided by considerations of outgroup comparison, reconstruction of an archetype, or to push the concept to its logical conclusion, without the consideration of manuscript authority of any kind. When we dip into a late and dependent manuscript for a uniquely preserved reading against the consensus of all other manuscripts, we are saying, with Bentley, "Reason and context are worth more than a hundred codices."

Furthermore, Dawe also must deny generally that scribes are capable of emendation (e.g., Dawe 1964, p. 106 on *Prom.* 562), and he must deny that true readings can crop up accidentally. If we find a satisfactory reading preserved in a single dependent manuscript, how can we really claim that the scribe was not at least as smart as we are and capable of inserting his own conjecture at this point so that the satisfactory reading is not part of the tradition in the strict sense at all? We assume too easily that these popular texts, often for use in schools, were only copied instead of being edited. Once deliberate editing of texts begins to replace simple copying, we no longer have straightforward descent of manuscripts, and cladistic methods no longer give a clear picture.

We have seen two ends of the spectrum: (1) cases where vertical transmission and uncontaminated text tradition makes the mechanical application of cladistic methods to reconstruct a single archetype a workable and successful method, with a claim to being scientific; and (2) cases of horizontal transmission, full of Byzantine, and even ancient, editing and conjecture. In the latter case, cladistic methods give little aid, and when we are deprived of a mechanical method, we rely more completely on reason and context, that is, the high degree of redundancy in the text, to reconstruct what would have been the author's words if they had been transmitted clearly; this latter has claims to being called an art. In any practical job of editing, however, both ends of this spectrum, and all intermediate stages, are likely to be brought into play.

But these theoretical difficulties about manuscript affiliation do not ultimately make much difference for the editor, because he can usually determine the polarity of the manuscript variants independently of the stemma. The chief practical use of stemmatic methods is to eliminate manuscripts from consideration. We have an advantage over the zoologist, who does not have the luxury of right animals and wrong animals, correct features and incorrect features. A derived character in a manuscript is always wrong and almost always bears a stigma of its wrongness. The redundancy of the text gives the critic an independent extra criterion for determining polarity. Furthermore, the task of editing a text is different from the task of establishing manuscript affiliations, which is only a preliminary and ancillary problem for the editor, who wants to establish the *ipsissima verba* of an ancient author and will use all available means.

When Norman Platnick, in the course of our collaboration, returned to me his revision of a section I had drafted, I found that at one point he had added to the perfectly reasonable word "stemma" the parenthetical explanation that it was an inverted cladogram. It struck me as curious that zoologists should live upside down in the droughty antipodes like this, and I began to reflect on why this should be. Conventions of expression tend to reflect the purposes behind them, and metaphors give clues to unspoken assumptions.

Zoologists really are most interested in the end-points of their trees, the individual taxa. They are interested in how real animals are related and the course of evolution that got them there. The more end-points, the more successful the analysis. They are excited about apomorphies.

The textual critic is not really interested in the end-points of the tree, that is, the specific manuscripts. Indeed, he tries to eliminate as many as possible. His only real interest is in reconstructing the archetype as a step in reaching the author's original text. Only those manuscripts that contribute information toward that goal are worthy of his notice. Apomorphies will all be eliminated from the reconstruction; he is interested in the purest plesiomorphy.

The result is that we put what we are most interested in at the top. The goal of the investigation is uppermost, the reconstructed plesiomorphic archetype for the textual critic, and the decisive autapomorphies for the zoologist.

For the zoologist the reconstructed ancestor is a fiction. It is a convenient way of presenting the organized information about the relationships of the real animals in question. Nobody tries to reconstruct a living, breathing thecodont or the protodipteron. But the reconstructed text of the textual critic is the real article, in a literal sense the text that Euripides wrote. Reconstruction is a serious business, and the only point of studying manuscripts at all.

The linguist seems to be the hybrid species in my classification. Like the zoologist, he is interested really in the end-points of the tree, and his reconstructions are also convenient formulas for presenting the organized information about the relationships of real languages. He does not reconstruct real languages, and we look with some amusement on Schleicher's famous attempt to write a fable in Indo-European. Nobody believes that if a flesh and blood Indo-European stood resurrected before us, with a turtle in one hand and a beech tree in the other, we could chat with him confidently.

But why then do linguists share the upside-down cladograms of the text critic? Here another deeply imbedded metaphor comes into play. For the zoologist the end-points of the diagram represent success, achievement, survival, and progress; each represents a triumphant adaptation with survival value. But the metaphor for the text critic and linguist is decay and corruption, a falling from perfection, descent. While modern sophisticated theory of linguistic change no longer puts it this way, the old myth of Babel and falling from grace lurks still in the orientation of their diagrams.

Notes

1. It is possible to see the beginnings of the method a generation earlier in Politian's analysis of the relationship of the manuscripts of Cicero's *Letters to His Friends*. In his *Miscellanea* (1489), Politian shows that MS Laurentianus 49.7 (P) is a copy of Laurentianus 49.9 (M) (*Misc.* chap. 18), and that P in turn is the archetype of a whole family of late manuscripts (*Misc.* chap. 25). See Timpanaro 1963, pp. 4–6; Reynolds and Wilson 1974, p. 128.

2. The direct evidence that this was the case is small but convincing. In Aristophanes' *Frogs* 52ff., Dionysus is lounging on shipboard reading a copy of Euripides' *Andromeda* (now lost). At *Frogs* 1114 the chorus says that there is no danger of the audience missing the parody because they are familiar with the originals from books. From Xenophon's *Anabasis* 7.5.14 we learn of a cargo of books lost in a shipwreck. A fragment of Eupolis (304 Koch) mentions Athenian bookshops. Plato's *Apology* 26D says that the works of Anaxagoras could be bought cheaply in the orchestra. Xenophon's *Memorabilia* 1.6.14 has Socrates reading and excerpting passages from books. See Schubart 1921; Kenyon 1951; Turner 1954.

3. The evidence comes from pseudo-Plutarch's *Lives of the Ten Orators* in the collection of Plutarch's writings called the *Moralia* 841 F.

4. The evidence comes from the works of the medical writer Galen 17.(1.)607.

5. We can often deduce when an error entered the tradition on the grounds that it depends on the misreading of a majuscule text rather than minuscule, or vice versa. On this transliteration or *metacharacterismus,* see Browning 1964. The Arabs learned paper making from Chinese prisoners taken at Samarkand in 751. See Reynolds and Wilson 1974, p. 52.

6. The archetype is the ancestor to all extant manuscripts and is reconstructed from their evidence. Note that the archetype is not the same as the *ipsissima verba* of the author, for the archetype may still contain errors that will be shared by all its descendants. Reconstructing the archetype is only a step in reconstructing the original text, which often requires that we deduce readings from the manuscript evidence which none of them have explicitly.

7. *Adagiorum Chiliades,* Basilea 1538, p. 209. This is the second edition of his *Adagia,* which he prepared on a visit to Italy in 1508 while he was a guest of Aldus Manutius in Venice (Sandys 1903–1908, 2:129). Timpanaro (1963, p. 8) says, "Ad ogni modo, veniva enunciato così da Erasmo il concetto di un capostipite perduto di tutti i codici giunti fino a noi, che permetterà di spiegare l'esistenza di errori comuni all'intera tradizione manoscritta."

8. *Catulli Tibulli Propertii nova editio,* Ios. Scaliger recensuit, Anversa 1584, p. 4.

9. This stemma is simplified by the omission of manuscripts H D O F, which do not change the situation appreciably, and I have left out of consideration Zuntz's argument (Zuntz 1965, p. 197) that there was actually an intermediate MS π between P and γ. I have worked out this stemma myself, but it is consistent with that of Turyn (1957, pp. 307–362, esp. p. 308) except that I have passed over the fact that *Med.* 1–231 does not belong to the Vatican class in A. The common errors at 1218 and 1359 demonstrate a "closed" tradition, that is, the manuscripts are descended from a single archetype. L P share many errors (e.g., 411) against A V B. P is independent of L at 1078 and 1158. L is independent of P at 282. Conjunctive errors for A V B are at

487, 816, and 837; conjunctive error for A V at 912; peculiar error of B at 981; peculiar error of V at 1191 and 1339; peculiar error of A at 1190.

10. Bentley said this in his 1711 edition of Horace on *Carm.* 3.27.15. If seen in context, the dictum sounds less radical, but it is often quoted unfairly as a slogan for the spirit of runaway conjecture. Bentley bears much of the blame, since his Horace is a notorious example of such speculative editing.

11. See the introduction to Hude's Oxford text, 2d ed. (1927), p. ix. "Codex P vero memoriam ita mixtam exhibet, vix ut origo eius certe definiri possit, et recte iudicasse videtur Kallenberg v.d. (in *Philol.* xliv, p. 740) huius codicis archetypum ad codices utriusque stirpis confectum esse." For the sake of providing a neat example, I have taken at face value the received judgment that P is a contaminated manuscript. I suspect the situation is much trickier, and a thorough cladistic analysis, for which there is no time here, would reveal that agreement between P and D indicates the reading of the archetype rather than contamination.

12. Wagner (1980, p. 181) deals explicitly with the cladistic problems that hybrid taxa present and concludes that all hybrid taxa should be removed from the data set used to construct the cladogram. I have been unable to take into account the recently published *Advances in Cladistics,* Volume 2, edited by Norman I. Platnick and V. A. Funk (New York: Columbia University Press, 1983), where papers by Farris, Wagner, Nelson, and others deal with issues of hybrid taxa, homoplasy, and reticulation.

13. See the introduction to Marchant's Oxford text of the *Cyropaedia,* p. v: "Si quis hodie indagare conatur quam ad rationem constituenda sit nova Cyropaediae editio, vix fieri potest quin in vasta variorum lectionum voragine fere submersus diu ac multum laboret."

14. Dawe (1964, p. 157) pays tribute to Pasquali, whose long, learned, and rambling book *Storia della tradizione e critica del testo,* 2d ed. (Florence: Le Monnier, 1952), criticized stemmatic theory and emphasized its shortcomings. For sensible remarks on the controversy, see Reynolds and Wilson (1974, pp. 212ff., 247–248).

References

Browning, R. 1964. "Byzantine Scholarship." *Past and Present* 28 : 3–20.

Dawe, R. D. 1964. *The Collation and Investigation of Manuscripts of Aeschylus.* Cambridge: Cambridge University Press.

———. 1973. *Studies on the Text of Sophocles.* 2 vols. Leiden: Brill.

Hall, F. W. 1913. *A Companion to Classical Texts.* Oxford: Clarendon Press. Reprint, Hildesheim, 1968.

Hennig, W. 1966. *Phylogenetic Systematics.* Urbana: University of Illinois Press.

Kenyon, F. G. 1951. *Books and Readers in Ancient Greece and Rome.* 2d ed. Oxford: Clarendon Press. (First edition 1932.)

Maas, P. 1958. *Textual Criticism.* Translated by Barbara Flower from the second (1949) edition of *Textkritik* 1927. Oxford: Clarendon Press.

Ostrogorsky, F. 1969. *History of the Byzantine State.* New Brunswick, N.J.: Rutgers University Press.

Page, D. 1934. *Actors' Interpolations in Greek Tragedy.* Oxford: Clarendon Press.

———. 1938. *Euripides Medea.* Oxford: Clarendon Press.

Platnick, N. I., and Cameron, H. D. 1977. "Cladistic Methods in Textual, Linguistic, and Phylogenetic Analysis." *Syst. Zool.* 26 : 380–385.

Reynolds, L. D., and Wilson, N. G. 1974. *Scribes and Scholars.* 2d ed. New York: Oxford University Press.

Roberts, C. H. 1954. "The Codex." *Proc. Brit. Acad.* 40:169–204.

Sabbadini, R. 1905–1914. *Le scoperte dei codici latini e greci ne'secoli XIV e XV.* 2 vols. Firenze: Sansoni. Reprint, 1967.

Sandys, J. E. 1903–1908. *A History of Classical Scholarship.* 3 vols. Cambridge: Cambridge University Press. Reprint, 1958.

Sauppe, H. 1841. *Epistola critica ad Godofredum Hermannum.* Lipsiae: Weidmann.

Schubart, W. 1921. *Das Buch bei den Griechen und Römern.* 2d ed. Berlin and Leipzig: De Gruyter.

Timpanaro, S. 1963. *La genesi del metodo del Lachmann.* Firenze: Le Monnier.

Turner, E. 1954. *Athenian Books in the Fifth and Fourth Centuries B.C.* London: Lewis.

Turyn, A. 1957. *The Byzantine Manuscript Tradition of the Tragedies of Euripides.* Illinois Studies in Language and Literature. 43. Urbana: University of Illinois Press.

Wagner, W. H., Jr. 1980. "Origin and Philosophy of the Groundplan-Divergence Method of Cladistics." *Syst. Bot.* 5:173–193.

Widmann, Hans. 1967. "Herstellung und Vertrieb des Buches in der griechisch-römischen Welt." Archiv für Geschichte des Buchwesens, Band VIII p. 545–640.

Wolf, F. A. 1795. *Prolegomena ad Homerum.* Halle: Orphanotropheum.

Zuntz. G. 1955. *The Political Plays of Euripides.* Manchester University Press.

———. 1965. *An Inquiry into the Transmission of the Plays of Euripides.* Cambridge: Cambridge University Press.

13

Representing Language Relationships

WILLIAM S.-Y. WANG

Efforts to define the relationships among the set of contemporary languages stem from a desire to know how they have derived from some other set of languages that were spoken earlier in time. In some complex way, the former set of languages has *changed* into the latter set. A true portrayal of such historical relationships must therefore rest on an accurate understanding of how such changes take place. As Hoenigswald remarks, "Much of the uncertainty in historical linguistics centers on the concept of the change event" (1973, p. x).

Change in a language takes place along many dimensions. It could occur in the ordering of words (syntax), in how morphemes form words (morphology), in the meanings of words (semantics), and in how the words are pronounced (phonology). My purpose here is to examine just one dimension of linguistic change: the way a phonological change spreads across the lexicon, or lexical diffusion, and some consequences of lexical diffusion on the portrayal of relationships.

An important key to understanding change has to do with how language is transmitted. Most of us have a limited ability to enrich or otherwise alter our linguistic behavior even after puberty, at least until the age of fifty or so, according to the Montreal study reported by Sankoff and Lessard (1975). But language change *within* individuals is minimal when compared with that *across* individuals. The major changes occur when linguistic traits are transmitted from one speaker to another. Work in historical linguistics has been primarily concerned with the transmission of first languages over centuries; less attention has been paid to the added complication of cross-linguistic interference.

Transmission of language occurs at two very distinct levels: one biological and the other cultural. We inherit the capacity for language, a capacity that is strikingly well developed in our species. This capacity includes all the motor, sensory, and cognitive equipment without which language cannot be learned and used. The major ingredients comprising this equipment evolved much prior to language, for purposes as fundamental as breathing, eating, and so on.

For this reason, phoneticians have long spoken of speech as an "overlaid" function, one that makes extensive novel use of preexisting organs such as the tongue or the jaw. By extension, we may think of language as "overlaid" as well, exploiting and reciprocally enriching powers of memory and reasoning in addition to the finely tuned perception and motor control that supports speech. The unique power of language is that it links and integrates all these preexisting components into one comprehensive symbolic system. Surrogates to speech, such as writing and signing, emerged considerably later.

Such a symbolic system must have conferred a tremendous adaptive advantage on our ancestors, both in enhancing the effectiveness of individuals and in coordinating activities of groups. It has been suggested that some parts of this equipment evolved along more specialized lines to favor the elaboration of language phylogenetically.

Drawing on the observation that creole languages are highly similar, even though they have arisen in different settings, Bickerton (1984) hypothesized that there is a genetically coded "language bioprogram" that is specific and exclusive for language. The bioprogram emerges in its clearest form, he argues, because the child is least guided (constrained?) by full adult language models in contexts where creoles arise. It is difficult to evaluate this hypothesis with confidence until further research can help us determine the extent to which the observation can be explained by the similar cultural demands made on the similar cognitive machinery that all humans share.

Included in the biologically transmitted global attributes that make language possible, there are also attributes of a much more local nature that influence individual linguistic behavior (see the discussions anthologized in Fillmore et al.) (1979). For instance, a recent paper by Smith et al. (1983) reports identification of a gene that plays a major etiologic role in one form of reading disability.[1]

Although exceptional forms of linguistic behavior such as dyslexia are easier to identify and investigate, it is likely that there are many other, more subtle individual linguistic attributes that are determined or at least strongly influenced by the genotype. It is not unreasonable to believe that factors such as the neuromuscular structure of the tongue or soft palate should have some bearing on the individual's speech habits. (Consider the remarkable feat we perform effortlessly in identifying a speaker after just a few words spoken over a telephone on the basis of voice attributes.) However, these attributes have not as yet been investigated from a systematic point of view.

Biologically, the capacity for language that we inherit may be likened to a mold whose detailed content is to be filled via cultural transmission. Such a hybrid arrangement gives language its flexibility so that it can keep pace with the ever-changing culture that it communicates.

A Melanesian child raised in Sweden will not acquire the fair hair or blue eyes of his or her neighbors, but will acquire a variety of Swedish that is not distinguishable from that of the neighbors. Hair and eye color are biologically transmitted, so the mode is strictly vertical, that is, from parent to offspring. The particulars of a language, as distinct from the capacity for language, are transmitted culturally. Here the transmission is no longer constrained to just the vertical mode.

A child learns at first from his immediate family, but the influence of the family is typically overtaken quite soon by other sources, such as that of the teacher and that of peers. These represent respectively transmission in the oblique mode and the horizontal mode (Cavalli-Sforza and Feldman 1981). Some of the sources may represent forms of speech that are quite different from those of the parents, for example, a playmate from a foreign country. The result of this multimodal transmission process is that the child will end up with a composite language uniquely his own.

To a certain extent, this ontogenetic picture has its counterpart in language histories. This parallelism was noted by Hugo Schuchardt in 1885, when he wrote, "Everything that holds true for the relationship between dialects on any level also holds true for the relationship between idiolects" (1885, p. 49). All too often the language of a speech community is a composite of various elements drawn from diverse chronological layers and linguistic sources. Unlike biological relationships, where parents are the exclusive donors of genetic material, this composite nature poses the greatest difficulty in applying tree diagrams to represent language histories.

English contains numerous traits that come from without. After its emergence as a result of the fusion of several Germanic languages from the north European coast, it borrowed extensively from Scandinavian in the ninth, tenth, and eleventh centuries and then even more heavily from Norman French for several centuries after that. But such a mosaic history is probably typical of all languages at one stage or another of their development. It is a natural outcome of the movement of peoples. The case of English is best known only because of the recency of this history and because of the current sociopolitical importance of the language.

Given that two languages are in close contact, we have no theory at present that will predict who will borrow what from whom. Sometimes a language brought in from outside becomes permanently established, in spite of the small number of speakers at the outset. This is the case of English in India, for example, which is even now expanding its influence. There are also cases where the conquerors give up their language in favor of that of their new home. This has happened repeatedly with the Altaic peoples in China: witness the Mongols of the Yuan dynasty and the Manchus of the Qing dynasty.

It appears that any linguistic trait is borrowable—from basic words, to morphological formation and syntactic order, to patterns of sound. One might think that words as basic as those for the low integers are not subject to borrowing, yet the Chinese numerals have been borrowed extensively into the neighboring languages of East Asia and Southeast Asia. We speak of "linguistic areas," when many traits spill across family lines from one neighboring language to another to form a typological group (cf. Masica). Southeast Asia, for example, is such an area, where neighboring languages of diverse sources share the trait of being "tone languages."

The abundance of language borrowings makes it extremely difficult to determine relationships on purely linguistic grounds, since it is sometimes impossible to know which shared traits are retentions from a common source and which are due to borrowing. However, there appear to be some constraints on borrowings, namely, which traits are more likely and which borrowings are contingent on which others. The systematic working out of these constraints has only recently begun. The results forthcoming from this research will be of great value toward understanding language relationships.

Against this composite background of language development, a truly complete history of a language must be immensely complicated. At any moment in time, a language may be likened to a mammoth tapestry woven from thousands of threads that lead back to numerous and diverse sources. Languages are related to each other according to the threads they may or may not share across time.

The complete construction of such language relationship networks is an extremely labor-intensive task that perhaps is now possible with the aid of powerful computer technology. In a sense, every time a new etymology is worked out or a new dialect atlas is compiled, we have traced a few more threads in the tapestry. Rather than depending on construction of the complete network, the task of the historical linguist is also to search for traits that may be better indicators (more diagnostic) of historical development than others.

Over the past decades, a variety of traits have been examined with different degrees of thoroughness for many language groups. We may think of such studies as producing a matrix where the columns are labeled by languages and the rows are labeled by traits. The cells can be filled by any of three values, a + if the trait in question is present, a − if the trait is absent, and a 0 if the trait is irrelevant.

We may then try to draw a tree from such a matrix. Each trait serves as a node on the tree where languages that have the trait branch one way and languages that do not have the trait branch the other way. If we are successful, this would result in a unique tree that portrays the historical development of the languages in the matrix. Often, however, this is not the case.

The typical problem that arises is the matrix that contains portions of the following form:

	A	B	C
x	+	+	−
y	−	+	+

That is to say, if we take x to be the determining trait then language B would group with A. On the other hand, if y were taken to be the determining trait then B would group with C.

Perhaps the most famous statement of this problem is that by Leonard Bloomfield (1933, p. 316) for the Indo-European languages. If the branching were determined by passive voice endings with r, then Italic would group with Celtic. If it were determined by feminine nouns with "masculine" suffixes, then Italic would group with Greek. If it were determined by the perfect tense being used as a general past tense, then Italic would group with Germanic. Conflicts such as this gave impetus, Bloomfield noted, to the formulation of the wave hypothesis, which takes into account the geographical distances among the languages.[2]

Some partial solutions to the conflicting branchings can be reached by resorting to some kind of ranking of the traits. One suggestion is that older traits should have precedence over younger traits, since the former reflect a chronologically prior change. As Watkins (1966, p. 32) put it, "The subgrouping question is partly one of relative chronology." In cases where the traits in question can be reliably dated, this suggestion is indeed helpful, as is shown, for example, in Ting's recent subgrouping of Chinese dialects, using documentary evidence. However, for most languages documents do not go back very far in time.

Another suggestion is to rank phonetically unnatural changes above natural ones. The natural changes are the dozen or so changes that recur frequently in the languages of the world, such as word final devoicing, palatalization before high front vowels, and vowel nasalization before nasal consonants. These changes take place again and again in distant parts of the world because

they are the global consequences of how we produce or perceive speech—
biological artifacts, in a sense. The reason behind the suggestion here is that it
is more likely for two languages independently to undergo the same natural
change than for them independently to undergo the same unnatural change.
Consequently, the sharing of unnatural changes is more diagnostic of shared
history.

Let us explore another direction in this general area, the possibility of
using individual words as linguistic traits in helping us establish linguistic relationships To do this, it is necessary to provide a few preliminary remarks on
the mechanisms of phonological change.

The mechanism some of us have been studying for the past dozen years
or so is called lexical diffusion. A collection of such studies has been anthologized recently (Wang 1977). The scenario suggested here is one in which
a change affects a few words at a time. Cumulatively through time, the change
diffuses across the entire lexicon. Three stages may be discerned in the process of diffusion: lagger words, which have not yet changed (the unchanged or
u-stage); words that exhibit alternative pronunciations (the variation or v-
stage); and the leader words, which have changed (the changed or c-stage).

	u	v	c
W1	W1		
W2		W2 ~ W2̲	
W3		W3 ~ W3̲	
W4			W4̲

In the example above, showing lexical diffusion operating on four words, W1
is the lagger, W2 and W3 are undergoing variation, and W4 is a leader that
has completed the change. A large body of evidence confirms this scenario in
a dozen or so languages. We are also beginning to understand how phonetic
forces interact with word frequency to determine which words will lead in the
change and which words will lag behind (Phillips 1983).

Suppose a language L split into L1 and L2 at time t. Suppose, further,
that a change C began to operate before t in L and continues to diffuse lexi-
cally in L1 and L2 after t. Since the words that change in L1 may not be the
same as those that change in L2, comparing the relevant words in the two languages would give us some indication of the degree of divergence between L1
and L2. We may generalize this reasoning as it applies to additional splits, say,
L1 into L1a and L1b. A synchronic study of how the relevant words overlap in
the three languages, L1a, L1b, and L2, should allow us to reconstruct the relative chronology of divergence among them. Following Hsieh (1973), I will
refer to such methods as diffusion overlapping.

These methods are similar to those of lexicostatistics in some ways. In lexicostatistic studies, a set of basic word meanings are taken as reference points, and the computation is done on how many of the words themselves are replaced in time. Such replacements typically take relatively long periods of time before useful figures can be obtained. In diffusion overlapping, on the other hand, the change in the phonetic form of the words is considered. Since this is generally a faster process, it is possible to compute relationships occurring in shorter time frames, say centuries (or decades?) instead of millennia.

Hsieh applied diffusion overlapping to a group of twenty Chinese dialects spoken around Shanghai, with respect to a sound change in lexical tone. Out of 533 words available to Hsieh, 490 are not diagnostic, since they have all changed in the same way in the twenty dialects. The remaining forty three words were used as criteria for grouping the dialects.

Using essentially the same method as that proposed by Hsieh, with minor modifications, Baron (1974) studied a group of eighteen Siyi dialects spoken near Canton. Here the change is in the vowel, from a high front /i/ to either a back or apical vowel. Even though the data in the Siyi case are limited to only nineteen diagnostic words, the results achieved are plausible.

The most extensive investigation of this sort is that reported by Krishna-murti and his collaborators (1983). The question they set out to answer is how to subgroup six Dravidian languages on the basis of a single sound change separating these languages from their common ancestor. In this study, they have the advantage of being able to use, as a point of departure, the *Dravidian Etymological Dictionary,* which contains the reconstructed words before the sound change in question began to operate differentially in the six languages. Sixty three words in their data have diagnostic value.

The number of possible subgroupings for even a small number of languages is astronomically large, a fact noted early in Greenberg (1957). The exact number depends on the types of tree diagrams allowed as well as on the number of languages to be subgrouped. The actual languages would correspond to the terminal nodes of the tree diagrams, while the nonterminal nodes would correspond to the ancestral languages posited for the various subsets of the actual languages.

In Meyers and Wang (1963), we distinguished between binary trees, in which every nonterminal node has exactly two branches, and normal trees, in which a nonterminal node has two or more branches. In both cases, the computation must be done recursively. Here I will state some of the results from that paper.

We will first derive the formula for $B(n)$, that is, the number of binary trees for subgrouping n languages. Suppose a binary tree with $n - 1$ terminal nodes is given, where $n > 1$. Then the total number of nodes in the tree is

$2n - 3$. Now a binary tree with n terminal nodes can be constructed from the old one by merely inserting a new node that dominates any one of the $2n - 3$ nodes and by hanging a branch from it. (It is irrelevant whether the new branch extends to the left or to the right from the new node.) Every binary tree with n terminal nodes can be obtained by repeating this process for all the $2n - 3$ nonterminal nodes. Thus, if there are $B(n - 1)$ binary trees with $n - 1$ terminal nodes, then there are $(2n - 3) \cdot B(n - 1)$ binary trees with n terminal nodes. Stated in formula, we have:

$$B_n = (2n - 3) \cdot B_{n-1}, \qquad B_1 = 1.$$

The computation of normal trees in which the nonterminal nodes may have more than two branches is more complex. Let us use the notation $N(n,i)$ for the number of normal trees with n terminal nodes and i nonterminal nodes. We can divide the recursion exhaustively to the two cases as follows.

Case 1: If a tree with $n - 1$ terminal nodes and i nonterminal nodes is given, then i trees, each with n terminal nodes and i nonterminal nodes, can be constructed from it merely by hanging the nth node onto a new branch issuing from any one of the i old nonterminal nodes.

Case 2: If a tree with $n - 1$ terminal nodes and $i - 1$ nonterminal nodes is given, then $[(n - 1) + (i - 1)]$ trees with n terminal nodes and i nonterminal nodes can be constructed from it in the following way. A new nonterminal node that dominates *any* one of the $[(n - 1) + (i - 1)]$ nodes of the given tree is inserted, and the nth terminal node is hung onto a branch from the new node.

We set up the initial value of $N(1,0) = 1$. Further, we rule out cases where the number of nonterminal nodes is not less than the number of terminal nodes. Then, $N(n)$ may be computed recursively by the following pair of formulas.

$$N_{n,i} = i \cdot N_{n-1,i} + (n - 1 + i - 1) \cdot N_{n-1,i-1}, \qquad N_n = \sum_{i=0} N_{n,i}.$$

Krishnamurti et al. (1983) make the assumption that the six Dravidian languages are related to each other in a strictly binary fashion. Using the formula developed above, we may compute $B(6)$, which turns out to be 945 binary trees. Their method evaluates each of these 945 trees against each of the sixty-three words with respect to the sound change in question.

As an illustration of their method, we may consider Figure 13.1. The upper case letters labeling the terminal nodes represent the six Dravidian languages to be subgrouped. The four tree diagrams are but a very small subset of the 945 trees to be evaluated for each of the sixty-three words. This figure deals with the word that appears as entry 4524 in the *Dravidian Etymological Dictionary* (Burrow and Emeneau 1961).

The lower case letters under the language labels show whether the word has undergone the sound change. Referring to tree 1, we see that the word is unchanged (u) in language G, omitted (o) in language K, and changed (c) in the remaining four languages. To account for this distribution by tree 1, we posit that there was a single change event that took place after G and K have split off from the others. In tree 1, the change is indicated with an x directly above the node that dominates the remaining four languages. To account for the same distribution of this word (*Dictionary* entry 4524), tree 2 would require four change events, and tree 3 or tree 4 would require two change events each. (See Figure 13.1.)

The final subgrouping is represented by the binary tree that gives the most plausible (i.e., simplest) history of changes for the largest subset of the sixty-three diagnostic words. The heartening result of their study is that the

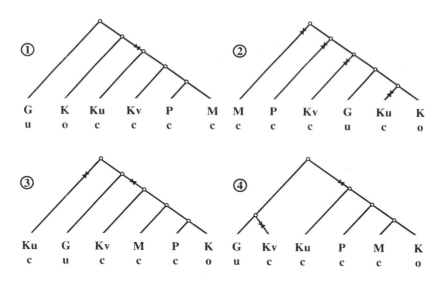

u = unchanged (cognate),   c = changed (cognate)

Figure 13.1. Reprinted from B. Krishnamurti, L. Moses, and D. Danforth, "Unchanged Cognates as a Criterion in Linguistic Subgrouping," *Language* 57 (1983): 545.

subgrouping that emerges coincides exactly with an earlier analysis (Krishnamurti 1978), even though the two studies are based on different types of information.[3] Even more strongly than the earlier investigations of Hsieh and Baron, and Dravidian study is a demonstration of the applicability of lexical diffusion to the analysis of language relationships. It is remarkable that, in all three studies, tracing the course of diffusion of a single sound change has revealed historical relationships that evolved over many centuries.

This direction of exploring language relationships is well worth pursuing, not only for more language groups but also for diverse types of sound change within the same group. At this initial stage of exploration many issues for which we have no answer are raised. Some of the problems that arise from the Dravidian study may become clearer as more cases are examined.

One problem has to do with the assumption necessary to the study that "a changed item will not revert to its unchanged form through a subsequent sound change" (Krishnamurti et al. 1983, p. 544). Yet, in addition to the cases the authors mention in their footnote 5, we can mention the report of Uriel Weinreich in which a Yiddish vowel first diphthongized and then remonophthongized; in fact, the title of his report is "A Retrograde Sound Shift in the Guise of a Survival." Such retrograde changes pose a serious problem for all of historical linguistics. It is not known at present how often such changes arise or how general they are.

A related issue is the decision of Krishnamurti et al. to treat variation words (i.e., words that exhibit both the unchanged form and the changed form) essentially as changed words. That is, their u/c is counted as a c. While this may be justified, as they do justify it, by disallowing retrograde changes, the information that the variation words contain (namely, that they represent a chronologically later layer of words to change) may be of some value in determining the subgrouping.

Another problem has to do with the assumption of binary splits. While it is plausible to argue that a community typically bifurcates when a subgroup moves away, the linguistic development may reflect relations that are more complex than binary ones. To take the example discussed earlier, note that in tree 1 the word has undergone change in each of the four languages Ku, Kv, P, and M. There is no basis for splitting them any further, as Krishnamurti was forced to do with the assumption of binary trees. A normal tree would allow all four languages to be dominated by the same node.

Working with binary trees is much easier than working with normal trees. For six languages, there are 945 binary trees, as evaluated in the Dravidian study, but there are 2,752 normal trees—almost three times as many (for seven languages, there are 10,395 binary trees, or 39,208 normal trees!). We need not only a sizable increase in computer time, but also richer data, to make the tree evaluation meaningful. Nonetheless, the results will ultimately be more realistic and significant in the sense that the method will not require us to posit more ancestral languages in the form of nonterminal nodes than is justified by the data.

Yet another problem has to do with the detection of borrowing between languages, i.e., horizontal transmission. While this problem is minimal in biological speciation (Feldman 1986), it is a central one in understanding language relationships. Methods need to be developed for quantifying the linguistic distances between languages. A possible approach would involve the comparison between the distance that would be theoretically predicted and the distance that is actually observed. If the observed distance is significantly less than the predicted distance, then we may be justified in explaining the difference between the distances as the result of borrowing. However, research along these lines has barely begun.

We have recently begun some work on data collected on vowel development in England. The data was reported by H. Orton in a series of publications dating back to 1962. Essentially, the observations were made at 311 sites in England, mainly on how various words are pronounced at these sites. The concepts that guide this work may be schematized in the table below.

	S1	S2	S3	S4	S5
$t0$	A(100)	A(100)	A(100)	A(100)	A(100)
$t1$	A(100	B(10)	B(50)	A(100)	A(100)
$t2$	B(10)	B(60)	B(90)	B(20)	A(100)
$t3$	B(50)	B(100)	C(10)	B(90)	A(100)
$t4$	B(100)	C(10)	C(80)	B(100)	A(100)

This table shows the hypothetical progression of a sound change across five sites (S1 through S5) during five points in time ($t0$ through $t4$). The notation A(100) indicates that the A pronunciation was used in 100% of the words. Thus we see that there is complete uniformity at $t0$, when all five sites showed 100 percent agreement. At $t1$, however, we see that the change has begun at S3, where 50% of the words now have the B pronunciation. It is an example of lexical diffusion caught at midstream in the operation. Furthermore, this change has spread to a neighboring site, S2, with a 10% B pronunciation. At $t2$ the B pronunciation has reached more sites, as well as a larger portion of the lexicon at these sites. Throughout the five points in time, however, S5 remains unaffected by this change.

We have applied these concepts to the development of Middle English i, and made a preliminary report of our efforts (Ogura, Wang, and Cavalli-Sforza 1986). Our hope is eventually to be able to translate the static pictures shown in dialect atlases into dynamic patterns that reveal the waves of propagation envisioned by the early linguists. The burgeoning field of computer graphics will be of great relevance toward this goal.

The preliminary results on Middle English *i* are encouraging. When we plot out the most frequent reflexes of this vowel on a map of England, there emerge various clines emanating from several known population centers. We can also see from the plots that there is a correlation between the distribution of the vowel reflexes and the rivers in the southern part of England. It appears from our data that where there are rivers, the similarity of the adjacent communities is decreased rather than increased, as had been reported for some other cases on the Continent. Clearly more data need to be examined before we can understand the nature of this contradiction.

At present we are adding the reflexes of several other Middle English vowels to the pool of data. When these are analyzed, we should be able to know the degree to which related sound changes conform to each other in geographical space. Linguists have often spoken of changes as being related to each other in the fashion of "push chains" or "drag chains," especially in situations where the Middle English vowels evolved their contemporary values. Looking at the problem from the point of view of geographical as well as phonological space may shed light on this issue.

The works discussed above approach the representation of language relationships in different ways, but they share the assumption that the differential rate of diffusion of a sound change in the lexicon can provide important insights on the operation of the change, and these insights must be incorporated in the representation of language relationships.

Acknowledgments

This chapter was written while I was a fellow at the Center for Advanced Study in the Behavioral Sciences. I would like to acknowledge the financial support of the Exxon Education Foundation and the Alfred P. Sloan Foundation (82-2-10), which facilitated its preparation.

Notes

1. This result was obtained by linkage analysis in nine families involving eighty-four individuals. Recent thinking has pointed to the possible hormonal basis for a variety of exceptional behaviors, including dyslexia. See the research news on Norman Geschwind in *Science* 222 (December 23, 1983), p. 1312.

2. These two hypotheses, likening language relationships to trees or waves, had their major proponents a century ago. Schleicher (1863), in using tree diagrams, was much encouraged by

the parallels he saw between linguistics and biology when Charles Darwin published the *Origin of Species* in 1859. For recent discussions on the relations between biological speciation and linguistic divergence, see the contribution by Konrad Koerner in this volume, as well as Feldman (1985).

Schleicher's student, J. Schmidt (1872), noted the intrinsic difficulties of tree diagrams and advocated wave diagrams to take into account the spatial dimension of language transmission. On the relative priority of Schuchardt (1885) versus Schmidt in this regard, see Malkiel (1955).

3. The sixty-three diagnostic words are listed in Krishnamurti et al. (1983, table 2), together with the change values. This makes it possible for the reader to experiment with various methods of subgrouping using the same data. It we treat o and u as equivalent, as the authors do in their procedure, then the following numbers emerge. Gondi is the most conservative language, with only two c's. Konda is the next conservative, with nine c's. The other four languages have many more changed cognates: Kui–53, Kuvi–50, Pengo–40, and Manda–30. We may further observe that these four languages show the same values in sixteen words and that in twenty-four words out of the sixty-three Kui and Kuvi show the same values and Pengo and Manda show the same values. These numbers would support the final result achieved by the authors. I thank L. Cavalli-Sforza for discussing with me the general problems of subgrouping.

References

Baron, S. P. 1974. "On Hsieh's New Method of Dialect Subgrouping." *Journal of Chinese Linguistics* 2:88–104.

Bickerton, D. 1984. "The Language Bioprogram Hypothesis." *Behavioral and Brain Sciences.* 7:173–221.

Bloomfield, L. 1933. *Language.* New York: Henry Holt.

Burrow, T., and Emeneau, M. B. 1961. *A Dravidian Etymological Dictionary.* 2d ed. New York: Oxford University Press.

Cavalli-Sforza, L. L., and Feldman, M. W. 1981. *Cultural Transmission and Evolution.* Princeton: Princeton University Press.

Cheng, C.-C. 1982. "A Quantification of Chinese Dialect Affinity." *Studies in the Linguistic Sciences* 12:29–47.

Cracraft, J. 1983. "Cladistic Analysis and Vicariance Biogeography." *American Scientist* 71:273–281.

Feldman, M. W. 1986. "On Speciation and Linguistic Divergence." In *Language Transmission and Change,* ed. W. S.-Y. Wang. Oxford: Basil Blackwell.

Ferguson, C. A., and Farwell, C. B. 1975. "Words and Sounds in Early Language Acquisition: English Initial Consonants in the First Fifty Words." *Language* 51: 419–430.

Fillmore, C. J., D. Kempler and W. S.-Y. Wang, editors. 1979. *Individual Differences in Language Ability and Language Behavior.* New York: Academic Press.

Greenberg, J. H. 1957. *Essays in Linguistics.* Chicago: University of Chicago Press.

Hoenigswald, H. M. 1973, *Studies in Formal Historical Linguistics.* Dordrecht: Reidel.

Hsieh, H.-I. 1973. "A New Method of Dialect Subgrouping." *Journal of Chinese Linguistics* 1:64–92.

Janson, T. 1977. "Reversed Lexical Diffusion and Lexical Split: Loss of -d in Stockholm." In *The Lexicon in Phonological Change,* ed. W. S.-Y. Wang, pp. 252–265. The Hague and Paris: Mouton.

Krishnamurti, B. 1978. "Areal and Lexical Diffusion of Sound Change: Evidence from Dravidian." *Language* 54:1–20.

Krishnamurti, B.; Moses, L.; and Danforth, D. 1983. "Unchanged Cognates as a Criterion in Linguistic Subgrouping." *Language* 59:541–568.

Labov, W. 1981. "Resolving the Neogrammarian Controversy." *Language* 57:267–308.

Malkiel, Y. 1955. "An Early Formulation of the Linguistic Wave Theory." *Romance Philology* 9:31.

Masica, C. P. 1975. *Defining a Linguistic Area: South Asia.* Chicago: University of Chicago Press.

Meyers, L. F., and Wang, W. S.-Y. 1963. "Tree Representations in Linguistics." In *Project on Linguistic Analysis,* Report No. 3, Ohio State University Research Foundation (N.S.F. Grant G-25055).

Ogura, M.; Wang, W. S.-Y.; and Cavalli-Sforza, L. L. 1987. "The Development of M.E. *i* in England: A Study in Dynamic Dialectology." In *New Ways of Analyzing Sound Change,* ed. P. Eckert. New York: Academic Press.

Phillips, B. 1983. "Word Frequency and the Actuation of Sound Change." *Language* 60:320–342.

Sankoff, D., and Lessard, R. 1975. "Vocabulary Richness: A Sociolinguistic Analysis." *Science* 190:689–690.

Schuchardt, H. 1885. "On Sound Laws: Against the Neogrammarians." In H. Schuchardt, *The Neogrammarians, and the Transformational Theory of Phonological Change,* ed. T. Vennemann and T. H. Wilbur Frankfurt am Main: 1972.

Schleicher, A. 1863. *Die Drawinische Theorie und die Sprachwissenschaft.* Weimar: Bohlau. 2d ed., 1873.

Schmidt, J. 1872. *Verwandtschaftsverhaltnesse der indogermanischen Sprachen.* Weimar: Bohlau.

Smith, S. D., et al. 1983. "Specific Reading Disability: Identification of an Inherited Form Through Linkage Analysis." *Science* 219:1345–1347.

Thomas, L. 1981. "Debating the Unknowable." *Atlantic Monthly,* July, p. 52.

Thomason, S. G. 1981. "Are There Linguistic Prerequisites for Contact-Induced Language Change?" Paper presented at the Tenth Annual University of Wisconsin-Milwaukee Linguistics Symposium.

Ting, P. 1982. "Phonological Features for Classification of Chinese Dialects." *Tsing Hua Journal* 14:257–273 (in Chinese with English summary).

Wang, W. S.-Y. (ed.). 1977. *The Lexicon in Phonological Change.* The Hague and Paris: Mouton.

———. 1979. "Language Change: A Lexical Perspective." *Annual Review of Anthropology* 8:353–371.

———. 1983a. "Explorations in Language Evolution." Hyderabad: Osmania University Press.

———. 1983b. "Variation and Selection in Language Change." *Bulletin of the Institute of History and Philology* 53:467–491.

———. (ed.). 1987. *Language Transmission and Change.* Oxford: Basil Blackwell.

Watkins, C. 1966. "Italo-Celtic Revisited." In *Ancient Indo-European Dialects,* ed. H. Birnbaum and J. Puhvel. Berkeley: University of California Press.

Weinreich, U. 1958. "A Retrograde Sound Shift in the Guise of a Survival." *Miscelanea homenaje a Andre Martinet.* Tenerife: Universidad de la Laguna.

14

Language Family Trees, Topological and Metrical

HENRY M. HOENIGSWALD

Historically, the tree schema to represent chronological relationship among languages is closely tied to the story of the Dispersal at Babel. Unlike its biological counterpart it never aroused resistance on the part of authority. Eric Hobsbawm has said that philology "was the first science which regarded evolution as its very core" and that it was further fortunate "because of all the social sciences it dealt not directly with human beings, who always resent the suggestion that their actions are determined by anything except their free choice, but with words, which do not" (Hobsbawm 1962, pp. 337–338). As a result, the "family tree" in historical and "comparative" linguistics became altogether part of the background. At first it was far too little questioned as such, and so were the entities for which its components stand: the "languages" or language states in their alleged internal homogeneity and synchronic purity, represented by the vertices, and the lines of "ancestry and descent," represented by the edges and paths. In fact, however, nineteenth-century linguistics, with its remarkable consistency and accomplishment, gave a new content to these schemas and terms. Not uncharacteristically, it turned out that the principles did not have to be explicitly clarified for the substantive work to bear fruit. When clarification *was* undertaken, this was often done with some recourse to biological evolutionism as it was understood and misunderstood by linguists. If we wish to understand language trees, it is at least as important to analyze positive practice (as laid down in the historical grammars and in the etymological dictionaries) as it would be to listen, prematurely as it were, to interpretations of the metaphor presented in an ideological or deductive vein.

One of the difficulties about the points and lines is that the entities they symbolize tend to be circularly defined and to lack easily recognizable physi-

cal boundaries. Unlike, say, the individuals in a family tree of the real (genealogical) sort, and, roughly, unlike species in large areas of biology, two
so-called dialects, for instance, may equally well figure as two separate objects or, alternatively, as one, having their differences ignored: dialects are not
necessarily discrete and will occur over a continuum of variability in space,
social structure, age structure, "style," and whatnot. Another difficulty has to
do with the idea of descent. It is customary to say that Modern English, or any
one variety of Modern English, is "descended from" (some variety of) Old
English—a language extant in written records—but has "borrowed from"
Norman French (likewise recorded). It is implied that "a language" cannot
have more than one "ancestor" (or "origin") unless those ancestors are in turn
related as ancestors and descendants—that, in other words, there is no
hybridization or "mixture" (Norman French and Old English are remotely related, but not related along one simple line of descent). It is difficult to clarify
this implication and to name the presumable criteria whereby descent and
borrowing are distinguished, although the distinction has seemed important
especially to those who wish to interpret linguistic relationship in terms of
general, extralinguistic history. Nor is the difficulty removed simply by saying
that a descendant or "later" stage of a language equals that language plus the
effects of a process called "change" which has affected it, so long as we do
not know how to recognize identity under change; there is no molecular per-
manence or genetic material to hold on to. To be concrete, it is not clear what
the difference is between the process whereby a "native" Old English ā (stān
'stone,' āþ 'oath') appears in Modern English as "long o"—presumably a
"sound-change"—and the process whereby the ü in French "loanwords" like
pure has been "adapted" to yū in the process of borrowing; both exhibit the
same "regularity." Similarly with changes that are not sound-changes: meat
narrows its meaning from "food" to "edible animal matter" as a matter of
native semantic change, but so does the English word "strike" narrow its
meaning when it is taken over into German, where it refers only (as Streik) to
a work stoppage.

If allowances are made, however, it is possible to agree on certain points.
It is known, for instance, that contact and imitation among and within speech
communities typically produce mergers of a phonological sort: individuals
who are not conditioned to hear a difference will not reproduce it. The merger
occurs only after the contact and may therefore be called an innovation.
Classical historical linguistics went far by concentrating on those "sound-
changes" that are instances of merger, and looked on merger as the prototypical innovation. This is because the converse is valid: whenever a
contrast in one language (e.g., -t- in German Vater vs. -d- in Bruder) is
matched by a noncontrast in the other (one and the same English -th- in both

"father" and "brother"), the presence of the contrast is to be recognized as a retention and its absence as an innovation (there is confirmatory evidence that the middle consonants in the two English words were still different in fairly recent times). Mergers are irreversible: whatever may at some future time befall the word-interior -*th*-'s of English cannot be expected to sort out such items as "father" and "brother" and "weather" (Ger. *Wetter*) and "whether" (Ger. *weder*) according to their antecedents. The common ancestor, in other words, was in this particular respect like German rather than like Modern English. Note that the question just what the sounds were in physical terms is not asked—only whether they were different (as in German) or identical (as in Modern English). When these observations and considerations are developed exhaustively into an application of the quaintly so-called comparative method, we obtain a collection of retentions, or a protolanguage, with certain testable and concretely interpretable properties. If applied, for instance, to the Romance languages, the result can be checked against the independent epigraphic or manuscript remains of classical Latin.

All innovations are not phonological—far from it. It is, however, chiefly the phonological innovations that may be recognized as innovations ipso facto (unlike, e.g., semantic shifts). Consequently, a line of descent can now be defined through a relational property: if two bodies of vocabulary (say, the "Anglo-Saxon" vocabulary of Modern English and the vocabulary of Old English, or the "French" vocabulary of Modern English and the vocabulary of Old French) are matched (and, in the simpler case, when the vocabularies of two languages, say, those of Italian and Latin, are matched) and if certain additional cautions are observed, it may be found that one body shows no innovations (i.e., homonymies) at all while the other does, in which case the former is the ancestor of the latter. If, on the other hand, both exhibit innovations (as would be the case, e.g., with Italian and Spanish), they are both descendants of a third—on record or not as the case may be, but in any event reconstructable.

This is how family trees are constructed. Note that what is observed are not simply similarities such as would perhaps be definable as shared traits. There are, of course, taxonomies of languages (called typologies) by traits (tone languages, word order types). The question is still whether there are nonarbitrary ways of selecting the traits. There is no doubt, in any event, that genealogical classifications and typological classifications may and often do conflict. Even in the artificially restricted field of phonological contrast, subfamilies are established not by shared *traits* but by shared *mergers*. Two languages, A and B, may share many retentions and therefore be quite similar but it may turn out that the few (and possibly hidden) innovations that are shared at all link A with C, and B with D:

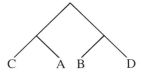

No statistical study of traits is required here. In fact, if one were gratuitously resorted to, it would be misleading.

The analogy with the stemmata of manuscript cladistics is obvious: recognizable innovations (i.e., homonymies due to merger) play the same role in language as copying errors, damage, and so on, do in manuscript history. One of the more interesting weaknesses of the analogy lies in the fact that in linguistics the relational criterion is strong, and the reasoning from content (say, the "naturalness" of a sound-change) is weak, while in textual stemmatics errors are in the main recognized qualitatively from an assumed knowledge of usage—although here, too, there have been interesting efforts to discover purely relational configurations of manuscripts and their variants.

Among the predictions and interpretations that these operations allow, the simplest is that if a language is discovered to be ancestral to another, its written records, if they exist, must be chronologically earlier (considering the maxim that "all languages change"). Other interpretations utilize identities among protolanguages arrived at from pairwise reconstructions of the type described above. These identities serve to organize simple lines of descent into trees. Three languages, A, B, and C, may thus behave as follows. (1) The three pairs (A/B, B/C, A/C) yield one and the same protolanguage, which is found to be (a) different from or (b) identical with one of the three given languages:

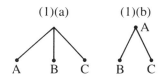

(2) Two of the three pairs yield one protolanguage, while the third pair yields a different one, in any one of the following four ways: (c) both protolanguages are different from any of the three given languages; (d) the two-pair protolanguage is identical with one given language, while the one-pair protolanguage is not identical with any given language; (e) the two-pair protolanguage is different from any of the given languages, while the one-pair protolanguage is identical

with one of them; or finally (f) the two-pair protolanguage is identical with one given language, and the one-pair protolanguage is identical with another given language:

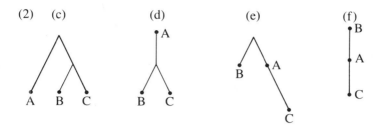

(2)   (c)   (d)   (e)   (f)

For a set of related languages to be amenable to arrangement in a tree, it would be necessary for all the smaller subsets sharing particular innovations to be enclosed within larger subsets sharing particular innovations. Suppose, for instance, that an ancestor, extant or reconstructed, possesses a fourfold phonological contrast,

$$t \quad d \quad s \quad z$$

and that there is a set A of descendant languages in which the contrast is reduced by merger, to wit

A d d

(i.e., proto-t and proto-d have merged into one and the same entity, here arbitrarily, if perhaps with a concern for phonetic concreteness, written d). Suppose further that there is also a set C in which it is s and z that are merged:

C s s

If A and C intersect in such a way that there are descendant languages

A ∩ C t t z z

a tree can be devised only if the same merger event is allowed to occur twice independently. If, however, A ∩ C = C, a tree

ancestor
|
A − C
|
C

would be appropriate to the data. In other words, to satisfy an orthodox tree, C must be a proper subset of A. Where it is in fact not a proper subset of A, the traditional remedies have varied. It may be that the lines of descent which by construction do not necessarily link whole languages but bodies of vocabulary within languages (see above) were not (perhaps because they could not be) properly separated ("borrowing" may not be distinguishable from "true descent" where the sources are not unrelated or remotely related languages like Old English and French but mere dialects of one another). It is in connection with such considerations that "clear cleavage" is insisted on as a prerequisite for the applicability of the tree schema and that in the absence of "clear cleavage" other geometric constructs (such as "waves") are proposed (see Bloomfield 1933, pp. 310–318). Attempts have been made to exclude or downgrade certain innovations as "trivial" and easily duplicated because of some universal or typological characteristic, and so on. A retreat into statistics has sometimes been tried but is notoriously difficult to reconcile with the framework we have set out so far—a framework aimed at distinguishing homonymy from nonhomonymy, but otherwise avoiding notational choices with statistical consequences of unknown magnitude.

It seems, however, that innovations are only selectively shared. In Finno-Ugric, Hungarian shares innovations with Ostyak but not with Finnish; in Indo-European, Celtic shares innovations with Germanic and other innovations with Italic (e.g., Latin), but none with Sanskrit. There are striking "similarities" between Latin and Sanskrit (see above), much prized precisely because they are retentions yielding precious antiquarian information with regard to the overall ancestor. It may be a general finding that, given three languages or three sets of languages, these may or may not be found linked somewhat randomly by shared innovations, but that some particular linkings are bound to be conspicuously missing where numbers greater than three are involved. This could reflect the circumstance that "borrowings" occur most often among geographic neighbors and that, for example, a two-dimensional arrangement like this

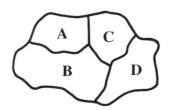

existing for a long period of time and not subject to upheavals would militate against A and D sharing innovations not also shared by B or C or both. The Indo-European sharings are among languages the speakers of which were still, relatively speaking, neighbors in historical times. The Hungarians and the Ostyaks of Central Siberia are not neighbors. This may be taken as an argument in favor of assuming a greater degree of mobility for the speakers of Finno-Ugric, and only slow (?), radial expansion for the speakers of Indo-European.

In sum, language family trees arrived at by the "comparative" method represent a fairly successful effort to convert the traditional metaphor of language descent by differentiation into the product of a formal procedure. This procedure applies to ideal situations characterized by "clear cleavage"—a limitation that we can well understand as we ponder the way the notion of ancestry-and-descent is obtained in the first place. "Comparative" trees are rooted trees, with the lapse of time represented by the direction away from the root point. They are, however, understood to be not metrical: the length of the edges and paths has no interpretation. What counts is only the way the vertices are connected, that is, the way ancestral vertices dominate descendant vertices.

It is also possible to consider trees characterized not only by their non-metrical properties but in addition by the length of their edges and paths (to be precise, by the projections of these edges and paths onto a vertical timescale) (Hoenigswald 1973, pp. 46–54, 61). A measure of justification for this lies in the suggestion that there may be a relation between a lapse of time and the extent of linguistic change which it produces. There has been a great deal of discussion about how to quantify effects of change. The most elaborate effort along the lines of finding a workable parameter has been Swadesh's lexicostatistics, with certain aspects of vocabulary replacement within some continuing semantic framework the subject of observation (Sankoff 1973a). Ignoring the merits and difficulties inherent in that particular approach, we shall instead assume that some acceptable index of diversity due to the passage of time has been identified, and explore possible ways to reconstruct time intervals on that hypothetical basis.

If, for instance, it were statistically valid and true to say that on the average a certain fraction (say, 80 percent) of some part of the vocabulary is retained after t years (say, 1000 years), and if it were also the case that after another t (1,000) years, the same portion (80 percent) of the vocabulary thus retained is retained in turn, the total retention after $2t$ years would be, under certain conditions, the square of the original fraction (64 percent). Of, say, 100 semantic slots, 64 are filled

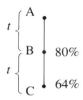

by the "same" item in languages A and C, while in the case of the remaining 36 meanings, A and C differ, each exhibiting a "different" item. It may be that some of the particular items in C, and none in A, are for some reason recognizable as innovated, but there is nothing in our numerical data alone to indicate direction in time. The finding is simply that C and A are removed from one another by a distance of 2t (we denote by t both the time and its measure in units of "distance").

Now consider an ancestor B with two separate descendants A and C. As the time t elapses, A and C retain the appropriate percentage of the vocabulary of B. If it is assumed for the sake of the argument that all items in the vocabulary in question (perhaps the "basic" vocabulary) are equally prone to replacement, the vocabulary that A and C still share would, in our particular example, once again be 64 percent of the vocabulary of B (Sankoff 1973a, p. 97). What matters is, in other words, the nonrooted, metrical tree.

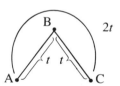

A given nonrooted, metrical tree containing three terminal vertices (three languages),

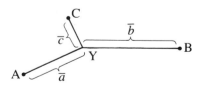

will, upon insertion of the root point X at one of three different locations (e.g., at A, at the node Y, or between A and Y)

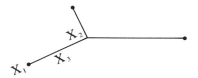

assume the following quite different shapes:

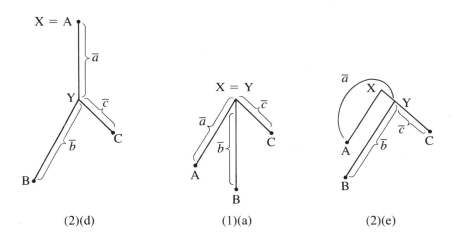

(2)(d)　　　　　　(1)(a)　　　　　　(2)(e)

where each of the distances $\overline{AB}$, $\overline{AC}$, $\overline{BC}$, as well as $\overline{AY}$ ("$\bar{a}$"), $\overline{BY}$ ("$\bar{b}$"), $\overline{CY}$ ("$\bar{c}$") remains nevertheless constant. The nonnoded shapes (1)(b) and (2)(e) and (f) (pp. 260–61) may be seen as showing Y = A ($\bar{a}$ = 0); the sub-ancestor is identical with one of the given languages. The (2)(c) shape above is, in fact, the nondegenerate, most general one.

If we express the "distance" of the root point (overall ancestor or proto-language) from A by $\bar{\xi}$, we may formulate certain relations among distances, tree shapes, and time depths (on the vertical scale) that are of interest. In view of the geometry of the (2)(c) tree shown in Figure 14.1, the following holds, for example:

(3) $\bar{a} = \frac{1}{2}(\overline{AB} + \overline{AC} - \overline{BC})$ (see Hoenigswald 1973, p. 48)

(4) $\bar{b} = \frac{1}{2}(\overline{AB} + \overline{BC} - \overline{AC})$

(5) $\bar{c} = \frac{1}{2}(\overline{AC} + \overline{BC} - \overline{AB})$

(6) X − A (i.e., the time depth of X, measured, in units of t, from A) = $\bar{\xi}$

(7) X − B = $\bar{a} + \bar{b} - \bar{\xi}$

$$(8) \quad X - C = \bar{a} + \bar{c} - \bar{\xi}$$
$$(9) \quad A - B = \bar{a} + \bar{b} - 2\bar{\xi}$$
$$(10) \quad A - C = \bar{a} + \bar{c} - 2\bar{\xi}$$
$$(11) \quad B - C = \bar{b} - \bar{c}$$
$$(12) \quad Y - A = \bar{a} - 2\bar{\xi}$$
$$(13) \quad Y - B = \bar{b}$$
$$(14) \quad Y - C = \bar{c}$$
$$(15) \quad X - Y = \bar{a} - \bar{\xi}$$

Expressions 9 to 12 may take positive and negative values, according to whether A (B, Y) is earlier or later than B, C (C, A). The ways certain inequalities in time depths can be accommodated by tree shapes has been partially explored elsewhere (Hoenigswald 1973, pp. 50–53). Where A, B, and C are contemporary to one another (e.g., where they can all be studied at present), the only possible shapes are (1)(a) and (2)(c)—the latter in all three

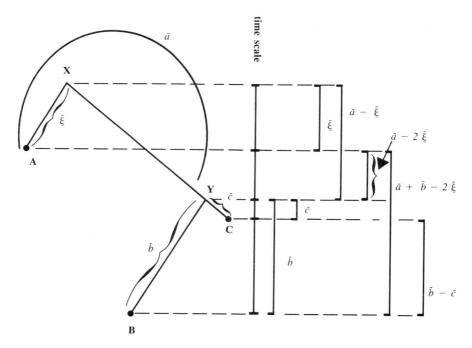

Figure 14.1.

of its varieties, that is, with A or B or C dominated immediately by X. Where two languages out of the three are contemporary and the third is earlier (say, preserved in written records), there are eight possible trees: three of shape (2)(c), two of shape (2)(e), and one each of shapes (1)(b) and (2)(d), as well as (1)(a). Or if A and B are contemporary and C is older, and $\overline{AC} + \overline{BC} = \overline{AB}$, as well as $\overline{AC} > \overline{BC}$, the tree is necessarily of shape (2)(e), with C occupying the interior position. And so on. It remains to state the appropriate theorems in a more general way (for n languages).

"Glottochronological" procedures of the kind described are resorted to in order to answer questions having to do with the subancestral structure of language families. In this respect, the metrical trees are expected to converge with the topological trees constructed by the comparative method (see the early part of this chapter). In addition, glottochronology claims to answer questions on the absolute antiquity of attested or reconstructed ancestral and subancestral languages.

References

Bloomfield, L. 1933. *Language*. New York: Holt.
Dobson, A. J. 1969. "Lexicostatistical Grouping." *Anthropological Linguistics* 11: 216–221.
Dyen, I. (ed.). 1973. *Lexicostatistics in Genetic Linguistics*. The Hague and Paris: Mouton.
Hobsbawm, E. 1962. *The Age of Revolution 1789–1848*. New York: World Publishing Co.; Weidenfeld & Nicholson.
Hoenigswald, H. M. 1973. *Studies in Formal Historical Linguistics*. Dordrecht: Reidel.
————. 1979. "Degrees of Genetic Relationship Among Languages." In *Commemoration Volume*, ed. S. P. Mallik, pp. 113–115. Burdwan, India.
Sankoff, D. 1973a. "Mathematical Developments in Lexicostatistic Theory." In *Current Trends in Linguistics*, ed. T. A. Sebeok, pp. 93–113. Paris: Mouton.
————. 1973b. "Parallels Between Genetics and Lexicostatistics." In *Lexicostatistics in Genetic Linguistics*, ed. I. Dyen, pp. 64–74. The Hague and Paris: Mouton.

15

Computational Complexity and Cladistics

DAVID SANKOFF

What cladistics gains in explanatory power over phenetics, it loses manyfold in mathematical and statistical tractability. More correctly, the global tree optimization procedures favored by cladists are necessarily far more complex than the stepwise agglomerative grouping methods traditional among pheneticists, though strictly speaking there is no theoretical reason why either school should be tied more than the other to a specific type of method.

In the agglomerative methods, a distance matrix (or, alternatively, a similarity matrix) is calculated as a first step, summarizing the overall differences (or resemblances) between all pairs of objects being studied. Successively larger groups are formed and combined, one at a time, through manipulations on the matrix, yielding a tree-like structure (dendrogram, phenogram, hierarchy, etc.). At each step, the group to be formed or combined is chosen according to some optimality criterion. This technique, which aims at a "good fit" between the matrix and the tree, is relatively rapid and can accommodate many objects since the computation time and space need not exceed a constant times the square of the number of objects and may often require less.

The other approach is to try to construct a tree that is the most likely according to some phylogenetically easily interpretable criterion applied to each character, feature, or dimension separately. The usual criterion is that of "parsimony" or "minimal mutations," which simply ensures that the evolutionary hypothesis implicit in the optimal tree involves as little as possible identical mutations at identical gene sites occurring repeatedly in different evolutionary lines. (It does not stem from a belief that "evolution takes the shortest path possible," as is sometimes suggested.)

The construction of the minimal mutation tree is far more difficult than stepwise clustering. The amount of computation time necessary tends to grow exponentially with the number of objects, so that even extremely sophisticated programming cannot accommodate large data sets.

It is worth repeating that the increased difficulty lies not in cladists' attention to each character separately, instead of to the fit between an overall distance matrix and a tree, but rather in their formulation in terms of a global optimum instead of the stepwise optimality used in the agglomerative schemes. Pheneticists who aim for globally optimal classifications must face the same sort of computational undertaking as cladists. The basis of the difficulty is that the search for an optimal tree allowing ancestral nodes (i.e., nodes other than those representing the given objects being classified) belongs to the class of problems known as NP-complete. Any problem in this class, which has been intensively studied by combinatorialists, operations researchers, and computer scientists for years, does not admit an efficient general solution (e.g., in linear, quadratic or other polynomial time) unless all NP-complete problems do. And none of them does, as far as is known, although this has not been proved either.

Unfortunately the tree-optimization problem has the same fatal allure for cladists as does the trisection of an angle with ruler and compass for geometry students. More than one mathematically gifted biologist has devoted far too much time trying to do what has defied the entire current generation of computational specialists.

If an NP-complete problem cannot be efficiently solved, in general, it can be circumvented.

A frequent approach has been to use stepwise and/or iterative algorithms that seek optimality by adding objects one by one in an optimal way to a partial tree, as in phenetics, or by adjusting a tree through local changes, each of which decreases the number of mutations implied. Neither of these approaches guarantees optimality, and experience shows that with reasonably large data sets they tend to produce a number of different trees, each depending on some arbitrary initial decision on the order in which the objects are incorporated into the tree.

We will return to this problem of a multiplicity of solutions later, but first we will describe an alternate approach, which guarantees the optimal solution at the expense of imposing a sufficient number of constraints on the problem. Such constraints may or may not be available in sufficient numbers for a given data set; this limits the applicability of the method, but to the extent that they are available, it becomes very efficient.

In the next section we describe the method, which will be illustrated in the ensuing three sections with studies of bacterial evolution based on RNA sequence data carried out in collaboration with R. J. Cedergren (Sankoff, Cedergren, and McKay 1982). Then we will discuss briefly the use of character state trees. In the final section we deal with the problem of multiple solutions.

Imposing Constraints

The idea behind the method is to assume that the optimal tree does not involve some thoroughly startling grouping, from the biological point of view, such as the grouping of some chloroplast RNA and some *Bacillus* RNA more closely than the chloroplast groups with other chloroplasts and the *Bacillus* with other *Bacilli*. The key to the procedure, then, is to impose restrictions on possible optimal trees, restrictions that simultaneously represent fairly certain biological knowledge and reflect clear-cut patterns in the data. For example, in the transfer RNA data to be discussed in the next section, it is biologically obvious, and unmistakable in the sequence similarities, that spinach chloroplast and bean chloroplast should be grouped together, and that *Euglena* chloroplast groups with these at a higher level, and that none of the other organisms is more closely related to any of these than they are to each other. If a restriction is imposed to the effect that the optimal tree must be consistent with this fact, the number of possible trees to be examined is still very large, but has been reduced to a fraction of what it was without this restriction. If enough such restrictions can be imposed, the tree optimization problem can always be reduced to manageable proportions.

Without going into technical details, we now turn to the search for the optimal tree. First, how is the set of all trees satisfying certain constraints generated? Each branch of a tree defines a partition of the objects into two groups, namely, those defined by removing the branch, thus decomposing the tree into two smaller trees. Indeed, any tree is equivalent to the set of two-way partitions defined by the set of its branches. Not all possible sets of partitions of the objects into two groups define a tree, however. For example, the two partitions (1,2) (3,4) and (1,3) (2,4) cannot both be in the same tree, as can be quickly verified. It is relatively easy to check any pair of two-way partitions and see if they are "compatible" for tree construction or not. Thus one way of generating all trees on a number of objects is to make a list of all possible two-way partitions, see which ones are compatible, and then enumerate all the combinations of compatible partitions.

The constraints we have been discussing can also be formulated in terms of partitions. For example, if the constraint is that certain objects must be more closely grouped together in the tree than any of them is with any other objects, this defines a two-way partition that must be present in every tree satisfying the constraint. This greatly reduces the number of possible combinations of partitions, since any combination not containing the constraint-induced partition is rejected.

A second aspect of the search procedure is the evaluation of the number of mutations involved in each tree satisfying the constraints, in order to find the tree that minimizes this quantity. This is a relatively rapid, though non-trivial, procedure based on the principle of dynamic programming (Sankoff and Rousseau 1975). It is carried out for each character separately and requires computational time proportional to the number of objects times the number of characters.

Phenylalanine Transfer RNA

We had nucleotide sequences of nine organisms for tRNA^{phe}. There are 660,032 different (rootless) trees based on nine objects. It would not be impossible to test them all, but this is a sizable undertaking even on a computer. Thus we imposed the constraints mentioned above on the chloroplast data, equivalent to two two-way partitions, plus the constraint that the two *Bacillus* sequences be grouped together, equivalent to one partition. The effect of these constraints is to reduce the computing time to about that necessary for searching for an optimal tree on six objects. Here there are only 236 possibilities, which represents a 2500-fold reduction.

In examining each tree for the minimal mutational history it implied for

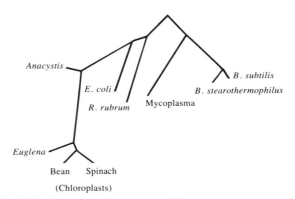

Figure 15.1. Minimal mutation tree for nine prokaryote phenylalanine tRNA sequences. Branch lengths proportional to weighted mutational distance between two end points.

the sequence data, we examined each of the ninety-odd sequence positions separately. We used a weighting function established earlier for the different types of mutation among the four nucleotides and for nucleotide insertion or deletion. The optimal tree (Figure 15.1) had length 35.68. Note that the position of the *root* is not determined by the optimality criterion. We have arbitrarily placed it to reflect the majority consensus on prokaryote evolution. Two trees that were almost as good (length 36.14) are schematized in Figure 15.2. From these results it follows that

1. *Anacystis* groups with the chloroplasts.
2. Mycoplasma groups with the *Bacilli*.
3. *Anacystis*-chloroplast does not group with *Bacilli*-mycoplasma.

5S RNA Data

For the nineteen 5S RNA sequences, the restrictions summarized in Figure 15.3 were assumed (cf. Hori and Osawa 1979, Fig. 4). As will be discussed later, the positions of *P. fluorescens, Clostridium* and *T. aquaticus*

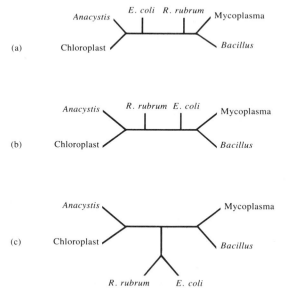

Figure 15.2. Schematic representation of minimal mutation tree (a), total length 35.68, and two next best trees (b) and (c), total length 36.14 each. Tree root and branch lengths not indicated.

have been placed elsewhere in some evolutionary theories. In these sequence data, however, the above subgroupings are all clearly indicated. In particular, the grouping of *T. aquaticus* with *Halobacter* seemed unavoidable in preliminary data manipulation, although the two sequences are not very similar.

The effect of these constraints is to cut the tree search down from an impossibly large number, on nineteen objects, to a very feasible task equivalent to searching the set of 2,752 trees on seven objects.

In studying 5S RNA, it is less clear than with tRNA which positions in one sequence correspond to which positions in the other. Many authors have studied this problem, which is essentially one of "character definition" (e.g., Sankoff, Cedergren, and Lapalme 1976). For the present study, we adopted an alignment of all nineteen sequences, representing a reconciliation of several previously published suggestions. The optimal tree, length 277.0, appears in Figure 15.4. The seven next best trees, with their lengths, are schematized in Figure 15.5 From these results the following seems clear:

1. Among the eubacteria, after the presumed remote origin of *T. aquaticus,* the *Bacilli* diverge from the rest as an early evolutionary event.

2. Probably *S. griseus,* but possibly *Mycobacter,* also branches off at an early stage.

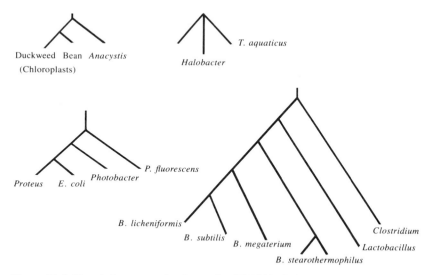

Figure 15.3. Restrictions on optimal trees for 5S RNA phylogeny. All trees tested must contain the above subtrees.

3. *R. rubrum* groups with the *Anacystis*-chloroplast group, or possibly with the enterobacteria.

4. Both the *Anacystis*-chloroplast group and *S. griseus* seem remote both from *E. coli* and from *Mycobacter.*

Given that these trees do not differ by more than a few mutations in their total cost, it is perhaps inappropriate to come to any more definitive conclusions. Nevertheless, among the required possible trees, the field of possible evolutionary hypotheses has been drastically narrowed. Note that the 5S RNA results confirm those of tRNA with respect to the remoteness of the *Bacillus* and *Anacystis*-chloroplast groups.

Evaluating Previous Hypotheses

The mutational cost criterion enables us to compare conflicting evolutionary theories. From the literature on prokaryote evolution, we collected four distinct evolutionary hypotheses (Schwartz and Dayhoff 1978; Hori and

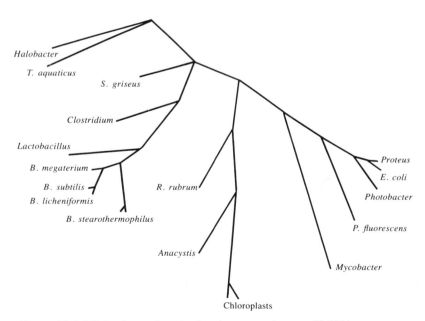

Figure 15.4. Minimal mutation tree for nineteen prokaryote 5S RNA sequences.

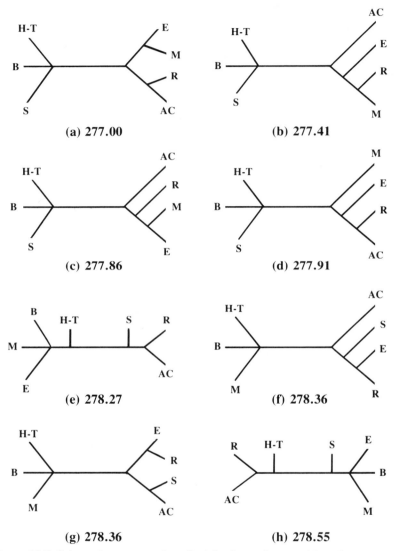

Figure 15.5. Schematic representation of minimal mutation tree (a), and seven next best trees (b)-(h). Numbers indicate total mutational (weighted) length. Key: H – *Halobacter;* T – *T. aquaticus;* B – *Bacilli;* S – *S. griseus;* E – enterobacteria; M – *Mycobacter;* R – *R. rubrum;* AC – *Anacystis* and chloroplasts.

Osawa 1979; Fox et al. 1980; and Küntzel, Heidrich, and Piechulla 1981). Since these did not all treat the same range of organisms, we enlarged the trees representing each theory by adding the missing organisms in a way that seemed to do the least violence to the authors' hypotheses. Because of this noncomparability, however, and because these theories do not necessarily represent their authors' current views, we refer to the corresponding trees as S, H, F, and K solely to indicate their ultimate origins in the literature. The trees are displayed in Figure 15.6, together with their lengths. It is clear that all are more costly than the tree we propose in Figure 15.4 and that the more they deviate from our tree the more costly they are.

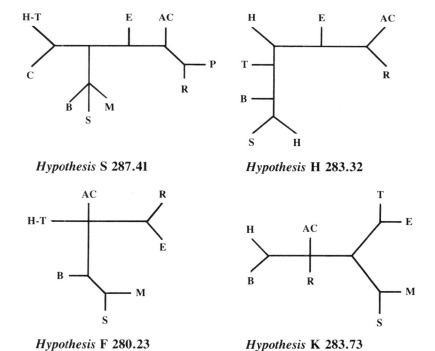

Hypothesis **S 287.41**   *Hypothesis* **H 283.32**

Hypothesis **F 280.23**   *Hypothesis* **K 283.73**

Figure 15.6. Evaluation of four previous evolutionary hypotheses. Numbers indicate total mutational (weighted) distance. Subtrees in Fig. 15.3 contained in all cases except where indicated. Key: H – *Halobacter;* T – *T. aquaticus;* B – *Bacilli;* S – *S. griseus;* E – enterobacteria; M – *Mycobacter;* R – *R. rubrum;* AC – *Anacystis* and chloroplasts; C – *Clostridium;* P – *P. fluorescens.*

Character State Trees

We have seen how *a priori* constraints on which objects must be close in the optimal tree render the cladistics problem feasible. In the RNA examples there were essentially five character states for each character, which we may represent by A, C, G, U, and ∅. Each of these may mutate into any of the others, though some types of mutations are more likely than others. In the more general cladistics problem, it is not true that any character may change into any other. A directed tree may sometimes be defined to indicate which changes are possible, as in Figure 15.7. Here a state *a* can give rise to state *b*, and *b* can give rise to *c* and *d,* but no other changes are possible. This type of information, where available, can be converted into constraints on the optimal tree, if we make the assumption of one-time-only mutations. It suffices to scan the set of objects with respect to the character in question, and to construct a two-way partition consisting of all those objects with character state *a* separated from those not having *a*. Another such partition may be constructed on the basis of state *c* and another for state *d,* If a number of partitions may be obtained from the states of each of several characters, the search for the optimal tree can be greatly narrowed.

Multiple Solutions

In agglomerative methods, different arbitrary choices that must be made during the procedure may produce somewhat different outputs. In global optimization methods, several solutions may be very close to the minimum, and we may not want to choose on the basis of very small differences. These and other situations may lead to more than one tree to account for the same data.

Two different trees may be very different, or they may be rather similar. In other words, the grouping information contained in one tree may be com-

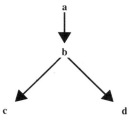

Figure 15.7. Hypothetical character state tree.

pletely different from, or very similar to, that contained in another. There is a simple way of measuring and extracting this common information. Consider the set of two-way partitions making up one tree and the set for the other. Those partitions that are in both sets represent their common element and define a new tree—the consensus tree (Bourque 1978). For example, the partitions defining the three trees in Figure 15.2 are:

(a) (AC) (ERMB)
    (ACE) (RMB)
    (ACER) (MB)

(b) (AC) (ERMB)
    (ACR) (EMB)
    (ACER) (MB)

(c) (AC) (ERMB)
    (ER) (ACMB)
    (ACER) (MB)

The consensus tree of any two and indeed all three of these is defined by the two partitions (AC) (ERMB) and (ACER) (MB), as depicted in Figure 15.8.

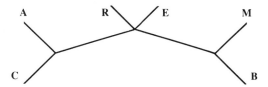

Figure 15.8. Consensus among three tRNA^phe trees.

Similarly, the consensus among the first four trees in Figure 15.5 is defined by the single partition (H-T, BS) (EMR, AC), as depicted in Figure 15.9. The tree in Figure 15.9 thus really contains only one grouping distinction. When a

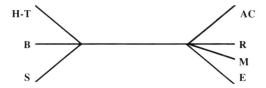

Figure 15.9. Consensus among 5S RNA trees.

set of trees contains no information in common, the consensus tree is just a "star," one central node with a separate branch for each object.

Conclusion

We have discussed a number of aspects of cladistic methodology arising from the NP-completeness of the unconstrained tree optimization problem. Clustering under constraints is a relatively new and promising approach and perhaps can contribute to making large-scale cladistic studies feasible.

References

Bourque, M. 1978. *Arbres de Steiner.* Ph.D. thesis, Université de Montreal.

Fox, G. E., et al. 1980. "The Phylogeny of Prokaryotes." *Science* 209:457–463.

Hori, H., and Osawa, S. 1979. "Evolutionary Change in 5S RNA Secondary Structure and a Phylogenetic Tree of 54 5S RNA Species." *Proc. Natl. Acad. Sci. (U.S.A.)* 76:381–385.

Küntzel, H.; Heidrich, M.; and Piechulla, B. 1981. "Phylogenetic Trees Derived from Bacterial, Cytosol and Organelle 5S RNA Sequences." *Nucl. Acids. Res.* 9:1451–1461.

Sankoff, D.; Cedergren, R. J.; and Lapalme, G. 1976. "Frequency of Insertion-Deletion, Transversion, and Transition in the Evolution of 5S Ribosomal RNA." *J. Mol. Evol.* 7:133–149.

Sankoff, D.; Cedergren, R. J.; and McKay, W. 1982. "A Strategy for Sequence Phylogeny Research." *Nucl. Acids. Res.* 10:421–431.

Sankoff, D., and Rousseau, P. 1975. "Locating the Vertices of a Steiner Tree in an Arbitrary Metric Space." *Mathematical Programming* 9:240–246.

Schwartz, R. M., and Dayhoff, M. O. 1978. "Origins of Prokaryotes, Eukaryotes, Mitochondria and Chloroplasts." *Science* 199:395–403.

INDEX

acmenism, 58
Adelung, Johann Christoph, 22, 24
adjacency, 75, 76
Adler, Mortimer, 55
Aeschylus, 227, 228, 232, 234, 237
affix, 17
Agassiz, Louis, 61, 144
agglomerative method, 269, 278
Albertus, Laurentius, 21
allele, allele frequency, 200
analogy, 88
ancestor, ancestry, 142, 143, 156, 157, 165,
    170, 217, 257–60
animals, 155, 157, 166
apomorph, apomorphy, 166, 182, 189, 218
Arabic, 9, 17, 19
Aramaic, 9, 19
archaic residue, 195
archetype, 60
Aristotle, 12, 29n15, 43, 54, 55, 65
Augustine, 12, 14–15
autapomorphy, 230, 239

Babel, Tower of, 8, 14–15
Barwick, Karl, 31n32
base, 6; base form, 7, 14
basic vocabulary, 264
Becker, Karl F., 58, 59
Bekker, Immanuel, 31n28, 235
Bengel, Johann Albrecht, 28n12
Bentley, Richard, 235, 236
Beowulf, 120
Berkeley, George, 40
Bible, 14–15
Bibliander, Theodor, 18–19

bilingualism, 224
biogenetic law, 125, 127, 138
Bloomfield, Leonard, 27n2
Blumenbach, Johann F., 73
Boas, Franz, 62
Bonnet, Charles, 71–73
Bopp, Franz, 7–8, 23–25, 56–58, 81–110
borrowing, 200, 222–24, 246, 258, 262
Borst, Arno, 5
botany, 26
branch length, 198
branching diagram, 5
Bréal, Michel, 59
Bredsdorff, Jakob H., 101n36
Buffon, Georges L. L., 73–76
Burgersdijck, Franco, 31n29
Buxtorf (Buxdorf), Johann, 29n18

Camper, Peter, 69
Cassirer, Ernst, 5, 101n38
Catullus, 231
cercopithecini, 198
chain (ladder), 50–54, 63, 67, 71, 72, 77;
    chain of being (*scala naturae*), 30n23
Chaldean. *See* Aramaic
Chambers, Robert, 26–27
change, linguistic, 43–45, 54, 87–92, 94,
    218, 219, 258; sound, 222, 243–55, 258,
    259
character: analysis, 156, 159–67, 204; defini-
    tion, 274; polarity, 141, 143; state tree,
    278; weighting, 164, 166, 189, 190
chemistry, 100n30
Chinese, 84
Chomsky, Noam, 27n2, 51

Chomsky, William, 30n25
Chrysippus, 12
Cicero, 236
cladistic analysis, 181–91, 194, 195, 198, 207, 219–26
cladogram, 142, 144, 147, 148, 156, 163, 167–74, 204, 210, 215
classification, 156, 158, 169, 217
clustering algorithm, 207
codex, 228
cognate: distribution, 216n2; word, 6, 209
Cohen, Morris R., 59
Collado, Diego, 21
commonality, 183
comparative anatomy, 61–63, 66, 69, 83, 93, 94
comparative method, 22–26, 61, 62, 66, 67, 219
compound word, 13, 19
Condorcet, Antoine, Marquis de, 51
contrast, phonological, 258, 259
contrastive linguistics, 62
convergence, 182, 189, 190, 222, 229; convergent trait, 195
core vocabulary, 209
correlation, 65
Cratylus, 5, 10, 25
creole, 224
cross-genetic comparison, 62, 71
Curtius, Georg, 7
Cuvier, Georges, Baron de, 37n79, 60, 61, 64–67, 88, 92, 94

Dahlmann, Hellfried, 31n32
Darmesteter, Arsène, 63
Darwin, Charles, 7–8, 25–26, 29n14, 36n78, 46–48, 51, 111, 115–19, 123–34, 156, 173
Darwin, George, 119
Daubenton, Louis J. M., 73
Daudin, Henri, 37nn80,82, 74
Dawe, R. D., 237, 238
De Sacy, Silvestre, 23, 25, 36n75
decline, declension, 12–13, 31n33
Demetrius, Triclinius, 228
dendrogram, 159, 161, 163. See also tree diagram
derive, derivation, 13–14, 17, 60–63
descent, descendant, 46, 143, 165, 170, 217, 257–60

development, 89, 91, 164, 171–74
diachrony, 57, 58, 63–65
Diderichsen, Paul, 37n79
Diderot, Denis, 69
Dietze, Joachim, 110
diffusion, lexical, 248–55; overlapping, 248
distance, 264–67; Manhattan, 200; measure, 198; patristic, 198; phenetic, 195, 198
distance Wagner technique, 200, 204, 210–14
divergence time, 194
Donati, Vitaliano, 26, 72
Donatus, Aelius, 13–15
Dreyfus, Hubert L., 37n88
Dubois, Jacques, 16
Duchesne, Antoine Nicolas, 111–12
Dwight, Theodore W., 117

electrophoretic data, 198, 204
eponym, 9
Erasmus, 227, 231, 236
errors: conjunctive, 229, 230, 236; in manuscripts, 227–40; separative, 230
Ethiopic, 19
etymology, 4–5, 7, 9–13, 15–16, 22–23, 31n29
Euripides, 227, 228, 232, 233, 239
evolution, 27, 155, 164, 166, 172–74; language, 126
evolutionary taxonomy, 219

Farris, J. S., 174, 200
Fechner, Gustav Th., 59
Ficino, Marsilio, 10
Fleisch, Henri, 27n2
fossil, fossil record, 139, 142–44, 158, 170, 174, 186–88, 194, 223, 224
Foucault, Michel, 37n88

Gabelentz, H. Georg C. von der, 91
Gellius, Aulus, 31n31
genetic locus, 198, 200, 204, 207
genetic distance. See distance
genotype, 218
Gesner, Conrad, 20, 70, 71
Giseke, Paulus D., 77
glottochronology, 194, 207, 209–14, 220, 267
Goethe, Johann W. von, 61, 69
Graff, Eberhard G., 87

Grammaire de Port Royal, 77n1, 103n45
grammar: comparative, 5, 23–25; traditional,
   4; universal, 23
Greek, 4, 9–12, 16
Greene, John C., 37n80
Grimm, Jacob, 7, 28n12, 82, 84, 98n15
Gudschinsky, Sara, 211

Haeckel, Ernst Heinrich, 47, 111, 123–34,
   188
Halhed, Nathaniel B., 67
Haller, Albrecht von, 72
Harris, Roy, 28n2
Hebrew, 4, 7, 9, 14–15, 17–21, 30n25
Hegel, Georg W. F., 59, 61
Helber, Sebastian, 21
Hennig, Willi, 142, 155, 158, 166, 170, 181,
   194, 229
Heraclitus, 47
Herder, Johann G., 82
Hermann, Johann, 73
Herodotus, 47, 52
Hesiod, 52
heterobathmy, 148
heterochrony, 189
Heyse, Karl, 100n30
hierarchy, 166, 181, 182, 185, 186, 225
His, Wilhelm, 134
historical linguistics, 194, 207–15, 246
historicism, 58
holophyly, 220
Holtzmann, Adolf, 101n30
Homer, 9
homology, 141, 145, 221–23, 225, 227, 231
homoplasy, 142, 156, 161–69, 189, 198
horizontal transmission, 237
Humboldt, Wilhelm von, 52, 82, 98n15,
   101nn30,38, 103n45
Hungarian, 20
Huxley, Thomas Henry, 66, 124
hybridization, 156, 165–74, 194, 223, 225;
   hybrid manuscript, 236

Ibn Quraysh, Yehuda, 30n25
India, 12
inflection, 12–14, 22, 83
ingroup. *See* outgroup
innovation, linguistic, 259–63
Isidore of Seville, 15, 20, 31n29

Islamic grammatical tradition, 17–18
Isocrates, 53

Jäger, Andreas, 45, 49, 50
Jakobson, Roman, 63
Japanese, 21–22
Jerome (Eusebius Hieronymus), 15, 32n44,
   32n45
Jones, Sir William, 66–67
Jussieu, Antoine Laurent de, 73
Justinian, 12

Kant, Immanuel, 68–70, 94, 95, 101n34,
   102nn40,43
Kempelin, Wolfgang von, 100n30
Kielmeyer, Karl F. von, 60
kinship, 118
Kiparsky, Paul, 86, 97n14
Kohlbrugge, J. H. F., 125
Krishnamurti, Bhadriraju, 249–53

Lachmann, Karl, 6, 28n12, 110–11, 236
Lamarck, Jean-Baptiste Pierre de Monet, 26,
   37n80, 73, 93, 111
language: classification, 259; law, 118–19;
   origin, 14–16
language tree, 257–67, 249–55; metrical,
   263–67; topological, 263–67
Latin, 9, 15–16, 21
Lebenskraft, 88
Leibniz, Gottfried Wilhelm, 22, 82
Lersch, Laurenz, 4
lexicostatistics, 194, 207–14, 263–67
Linnaeus, Carolus (Carl von Linné), 26–27,
   72, 73, 76, 77
loanword, 258
Lovejoy, Arthur, 50
Lull, Ramon, 29n15
Lyell, Sir Charles, 48, 55
Lyons, John, 4, 27n2
Lysias, 230

Maas, Paul, 229
Madvig, Johan Nicolai, 31n29, 58, 236
Maher, J. P., 111
Maine, Sir Henry Summers, 117–18
Malayo-Polynesian, 91, 93, 101n37
Manhattan distance. *See* distance
Manuel II, 29n15

Manutius, Aldus, 235
Masoretes, 18
Mayr, Ernst, 156
Meckel, Johann F., 60
Meillet, Antoine, 5, 28n11, 96n2
merger, phonological, 258, 259
metaphor, 40, 41, 46, 49, 52–55
metricity, 200
Mette, Hans Joachim, 31n32
molecular clock, 194
Monboddo, Lord (James Burnett), 67
monophyly, 169, 183
Morison, Robert, 27, 37nn85,87
morphocline, 144, 186
morphology, morphological, 7; kernel, 195;
    morphological system, 87
mother language, 20–21
Müller, Friedrich, 59, 129
Müller, Max, 35n73, 118–19
multilingualism, 224
multistate characters, 186
mutation, 218

natural history, 25
Nebrija, Antonio de, 9
Nei, Masatoshi, 200; Nei's standard genetic
    distance, 200
neoplatonism, 55, 61
Neophron, 234
NP-complete problem, 270
Nüsse, Heinrich, 55, 101n31

O'Cuív, Brían, 33n53
Old World monkeys, 200–03
ontogeny, 186, 188, 189; ontogenetic criteria,
    141
organic, organism, 7–8, 23, 41–43, 57, 62,
    81–107
Orton, Harold, 253
outgroup, outgroup analysis, 141, 156, 159,
    161, 162, 172, 183, 187, 200, 204, 229,
    233
Owen, Sir Richard, 61

paedomorphosis, 188, 189
Pagliaro, Antonino, 30n27
Palaemon. See Remmius Palaemon
paleontological rule, 187
Pallas, Peter Simon, 26, 37n82, 70–73, 112

parallelism, 162, 229
paraphyly, 143, 158
Paris, Gaston, 59
parsimony, 140–42, 159, 161, 164, 167,
    168, 182, 183, 185, 189, 204, 210
partition, compatible, 271–72
Pasquali, Giorgio, 28n12
Pastrana, Iohannes de, 29n15
pattern, 155, 156, 170, 173
Paul, Hermann, 29n16
pedigree, 123, 126
Peirce, Charles S., 59
Peter of Poitiers, 29n15
phenetics, 140, 143, 174, 219, 220. See also
    distance
phenogram, 198, 200, 204
phenotype, 218
philology, 110–12
phylogeny, 155, 156, 159, 166, 169–74; phy-
    logenetic reconstruction, 193
Piccolomini, Aeneas Sylvius, 20
Pictet, Adolphe, 47, 48
Pictet, François J., 48
pictorial illustration, 125, 126, 129–34
pidgin, 224
plant, 155, 156, 159, 166, 167, 171, 174
Platnick, Norman, 227, 239
Plato, 5, 10–12, 25, 30n27, 31n28, 43, 52,
    53
plesiomorph, plesiomorphy, 182, 183, 218,
    231
Plotinus, 61
polarity, 159, 162, 232, 238; of characters,
    see character
Politian, 231, 236
Porphyry, 6, 27, 29n15; tree of, 43
Port Royal, 77n1
Postel, Guillaume, 19
Pott, August F., 56, 59
prefix, 13, 17–19, 23
primitive, primitivum, 18, 159, 171
primitivism, 57
Priscian, 13–14
Proclus, 61
progressivism, 51, 52, 57
protolanguage, 15, 29n10, 221, 259, 260,
    265
prototype, 60
punctuated equilibria model, 215

Punic, 19
Pythagoras, 9

Qimḥi, David, 18
Qimḥi, Moses, 17
Quechua, 22

Rabinow, Paul, 37n38
races of man, 126
radiation, 63
Ranke, Leopold von, 116–17
Rask, Rasmus, 26
rate constancy, 200, 212, 215
Ray, John, 27, 37n86
recapitulation, 125, 127, 188, 189
reconstruction, 85, 87, 217
regressivism, 52
relatedness, 5–6, 9, 19–20, 91
relationship: cladistic, 155, 161–70; evolu-
    tionary, 155, 164, 166, 172, 173;
    hierarchical, 166; reticulate, 155, 157,
    167. *See also specific terms*
Remmius Palaemon, Quintus, 12
resemblance (generic, convergent, genetic,
    diffusional), 195
retention, linguistic, 259
reticulate, reticulation, 167, 169, 174, 225
retrograde evolution, manuscript analogue,
    229
Reuchlin, Iohannes, 18, 33n53
Richards, I. A., 40
Ritschl, Friedrich, 110–12
Rivinus, Augustus Quirinus, 27, 37n87
romanticism, 82, 84, 91, 93
root, 6, 21–24, 83–85
Rorty, R. M., 30n27
Ross, Alan S. C., 28n10
Rotta, Paolo, 5
Rousseau, Jean Jacques, 42
Royce, Josiah, 59
Rudbeck, Olaf, 71

Saadya, Gaon, 17
Sanskrit, 4
Santo Tomás, Domingo de, 22, 34n70
Sapir, Edward, 195
Saumaise, Claude, 45
Sauppe, Hermann, 230
Saussure, Ferdinand de, 3, 27n1, 48, 103n46

Savigny, Friedrich Carl von, 102n40, 116–17
Scaliger, Joseph Justus, 20–22, 227, 231,
    236
Scaliger, Julius Caesar, 16
Schlanger, Judith, 86, 87, 99n17
Schlegel, August Wilhelm, 8, 23, 25, 35n74,
    52, 55, 61
Schlegel, Friedrich, 7–8, 30n23, 55, 64, 66,
    67, 81–85, 90, 99n18
Schleicher, August, 5–7, 25, 26, 24n14,
    36n78, 46–48, 52, 56, 96, 109–12, 119,
    126, 239
Schleiden, Matthias Jacob, 36n78
Schmidt, Johannes, 29n16, 110
Schmidt, Rudolf, 32n46
Schuchardt, Hugo, 29n16, 245
sequence, 272–75
shared innovation, 220, 227, 229
similarity measure, 209
Smith, Adam, 42
Socrates, 10–11
Sophocles, 228, 237
species, 218
Sperber, Alexander, 27n2
Spitzer, Leo, 55
stage, linguistic, 258
Stam, James, 5, 28n8
standard language, 221
Steinthal, Heyman, 5, 56, 59
stem, 6–7
stemma, stemmatics, 110–11, 116, 229–40,
    260
Stiernhielm, Georg, 45
Stoicism, 12, 15, 16, 24, 32nn46,47
stratigraphy, 143, 144, 147, 149, 187, 188
suffix, 13, 17–19, 23
superorganism, 59
synchrony, 57, 58, 63
Swadesh, Morris, 194, 213, 263
symplesiomorphy, 141, 195
synapomorphy, 141, 144, 163, 183, 190,
    195, 230
systematics, 164; cladistic, 158, 162–70;
    evolutionary, 157

Talmud, 15
taxonomy, 26
Tene, David, 30n25
tenor, 40, 51

theme, 6–7, 29n18
Thompson, Sir D'Arcy W., 69
time depth, 248–55, 265; formula, 216n2
Timpanaro, Sebastiano, 28n12, 235
transformation series, 227, 233
tree, 40, 50–52, 54, 63, 74, 125–27, 133–34; evolutionary, 123; genealogical, 124; oak, 127, 133. *See also* language tree
tree diagram, 5–6, 26–27, 28n12, 110–12, 115, 141, 142, 157, 158, 159, 163, 194, 200, 229–40, 257–67, 269–80. *See also* language tree; dendrogram
tree optimization, 269–80
triangle inequality, 200, 216n1
Turgot, A.-R. J., 101n36
typology: of errors, 232; of language, 23–25, 221, 259

Ullendorff, Edward, 32n43
UPGMA clustering, 210
Uschmann, Georg, 111

Valla, Lorenzo, 12, 31n31
variation, 221, 225
Varro, Marcus Terentius, 12–13
Verburg, Pieter, 86

Veronese, Guarino, 9
vertical transmission, 238
Vicq-d'Azyr, Félix, 73, 77
Voegelin, Charles F., 71
Von Baer, Karl Ernst, 61, 188
vowel alternation, 90

Wachter, Johann Georg, 22
Wagner, H. W., 156, 158
Wedgewood, Hensleigh, 119
weighting: of characters, *see* characters; of features, 229
Whitney, William Dwight, 42, 56, 59, 118–19
Wittgenstein, Ludwig, 47
Wohllautgesetze, 87, 88
Wolf, Friedrich August, 23, 235
word: origin, 10–11; primordial, 24–25
written records, 221, 223

Xenophon, 53, 237

Zergliederung, 85, 95
Zuntz, Günther, 232–33
Zwirner, Eberhard/Kurt, 100n30